Elevators

ELEVATORS

ELECTRIC AND ELECTROHYDRAULIC ELEVATORS,
ESCALATORS, MOVING SIDEWALKS,
AND RAMPS

by F. A. Annett

Third Edition
290 Illustrations

McGRAW-HILL BOOK COMPANY
New York   Toronto   London
1960

ELEVATORS

Preface

Since the second edition of this book was published in 1935, the applications and operation of vertical-transportation equipment and moving sidewalks have been practically revolutionized. Among the many developments are:

Automatic dispatching of banks of passenger elevators according to changes in building traffic pattern; attendantless operation of passenger elevators in office and other heavy-traffic buildings; extensive use of electrohydraulic elevators and moving stairways; the application of moving sidewalks and ramps for mass movement of people; special designs of automatic garage elevators; and other innovations.

Some of these systems cannot be classified as electric elevators, but they should nonetheless be included in any treatment of elevators. To do this, the book's title has been changed from "Electric Elevators" to "Elevators," with the subtitle Electric and Electrohydraulic Elevators, Escalators, Moving Sidewalks, and Ramps.

The text has been revised and new subjects have been added to bring it up to date and to comply with the 1955 American Standard Safety Code for Elevators, Dumb-waiters and Escalators. Old Chapters 1 and 3 have been consolidated into one, with new material added, including information on manlifts, screw lifts, freight elevators, garage automatic-parking elevators, home elevators, and stairlifts. The resulting new Chapter 1 gives a broad general understanding of d-c and a-c powered elevators.

Old Chapter 2, dealing with double-deck elevator cars and two elevators operating in one hoistway, has been omitted. When the second edition was published, such elevators had been developed to meet the needs of proposed very high office buildings. These fell foul of economic laws, and with them, up to now, the elevators intended for them. If any architect follows Frank Lloyd Wright's idea of building a mile-high skyscraper, then we may have to dust off old Chapter 2.

Old Chapters 6 on d-c brakes and 7 on a-c brakes are rewritten in

v

one (Chapter 4). The material on elevator-motor controllers has been considerably reduced, and a new chapter added on d-c controls for a-c elevator motors. These controls, now used extensively, have d-c magnets and brakes to eliminate noise and other objections to a-c designs.

Attendantless operated elevators, automatically dispatched according to the changing daily traffic patterns in busy office buildings, are one of the marvels of modern vertical transportation. These have been comprehensively treated in a new chapter in which two different systems and the equipment that makes possible their safe and efficient operation are described.

In Chapter 11 on door operators new material has been added on freight-elevator doors and their operators. Chapter 12 on signal systems now includes new material on equipment that shows waiting passengers where the cars are and the direction in which they are moving, on electronic push buttons, and on car-position indicators.

Hydraulic elevators and lifts, each with its own motor-driven pump and electric control, form an important segment of passenger and freight vertical transportation systems for low and medium rises. These are covered in two new chapters, one on the elevators and their hydraulic equipment and the other on their electric controls.

Because people have become so convenience-minded, moving stairways (escalators) have become a part of the vertical transportation system of many public buildings. For safe mass movement of people horizontally and up or down inclines, moving sidewalks and ramps, even though newcomers, are rapidly finding ever-increasing applications. These methods of low-rise vertical transportation systems have a chapter devoted to their construction, operation, and control.

Of the total text over 150 pages is new material, much of which has appeared in *Power*, under the author's name. The practical treatment found in the two previous editions is featured in this one, making it suitable for all electrical workers interested in the construction, operation, and maintenance of vertical transportation equipment. The book will also provide vocational-school students with an intimate knowledge of all types of elevators.

In a practical book of this kind, a large part of which deals with types of equipment available, the author must lean heavily on the generosity of its builders for data, photographs, and drawings. Appreciation is here expressed to the many companies and others who have helped to make this edition possible. Credit for specific data and other material is given in the various chapters.

F. A. Annett

Contents

CHAPTER 1

Types of Machines—Direct Current and Alternating Current

Historical Sketch. Elevators, even when defined as devices for vertically lifting persons and materials, are by no means a modern invention. It is a matter of record that Archimedes, more than two hundred years before the Christian era, built an elevator. This device, although operated by manpower, was not dissimilar in principle to present-day drum-type electric elevators, in that it had a drum on which a hoisting rope was wound. In the latter part of the seventh century a hydraulic-type elevator was introduced into England, but it is only since 1850 that real progress has been made in elevator development. During this period four types of elevators have been developed: hydraulic, steam, electric, and electrohydraulic. Only electric and electrohydraulic designs are now in general use.

A drum-type elevator consists of a car attached by one or more ropes to a winding drum to which a counterweight is generally connected (Fig. 1-1). This drum may be electric-motor driven through a worm and gear, a spur gear, a combination of both, or in combination with one or more belts. The drum has grooves in which the ropes run with counterweight ropes connected at one end and car ropes at the other. Both sets of ropes run in the same grooves, counterweight ropes unwinding when car ropes wind and vice versa.

An earlier type of elevator that still survives in some old buildings operates with straight and crossed belts (Fig. 1-2). The motor is belted to a line shaft, which has a wide pulley on it, connected to idler pulleys A and B on the elevator-machine worm shaft by a straight and a crossed belt. Either the motor is started by hand and allowed to run continuously or it is started and stopped from the elevator platform by pulling on a rope. After the motor is running, the elevator is started by pulling on the operating rope in the car, which in turn operates a belt-shifting mechanism through the shipper wheel S. This shifts either the crossed or straight belt to the driving pulley C, which is keyed to the worm shaft of the machine. Shifting the straight belt to the driving pulley gives one direction of motion to the car, while shifting the crossed belt to pulley C imparts the opposite direction. With this type of machine

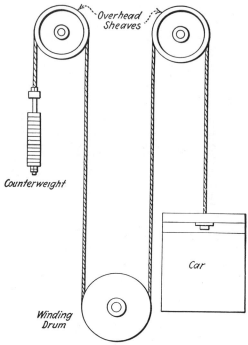

Fig. 1-1. Diagram of how car and counterweight are connected to the winding drum of a drum-type elevator machine.

the motor always runs in the same direction. A great many different arrangements of this type of machine have been developed, both for operation from the power line shaft along with the rest of the machines in the building and from an individual motor, as explained in the foregoing, but they all involve the same principle.

The Warsaw Elevator Company's machine (Fig. 1-2), is arranged for ceiling mounting, while the type shown in Fig. 1-3 is intended for floor mounting. Figure 1-3 indicates clearly the belt-shifting mechanism and how it operates from the shipper wheels S. Rollers O and O' are attached to the shipper-wheel rim and mesh into the forked castings G and G'. Casting G is attached to rod J and G' to

J' by setscrews. When the shipper wheel is turned in a clockwise direction, roller O' engages the fork G', moving it—and with it rod J' and the belt shifter K'—to the left. The belt shifter moves the belt off idler A onto tight pulley C and starts the elevator machine in a direction corresponding to the motion of the shipper wheel. When the operating rope is pulled to its central position to stop the car, this brings the shipper wheel to neutral position, as shown in Fig. 1-3. Pulling the operating rope so as to turn the shipper wheel in a counterclockwise direction would cause roller O to move fork G to the right. Fork G would take belt shifter K with it, which would move the belt off idler B onto the tight pulley C and impart opposite direction of motion to the elevator.

Fig. 1-2.   Belted type of elevator machine for ceiling mounting.

Improved Drum Types.   A step in the direction of improving drum-type elevators was to do away with the line shaft with its straight and crossed belts and to belt the motor directly to the driving pulley on the worm shaft. This involved a control for the motor that not only started and stopped it automatically, but also reversed it so as to give the elevator a direction of motion corresponding to the position to which the operating rope through the car was pulled. The same principles that prevailed in this belted-type elevator prevail in the modern drum-type machine. In the latter type the motor, winding drum, worm, and gear are mounted on the same bedplate, with the motor coupled directly to the worm shaft (Fig. 1-4).

Elevator-machine controls may be grouped into two general classes: semimagnetic and full magnetic. Semimagnetically controlled machines are operated by a rope running through the car, by a lever, or by a handwheel attached to the operating rope. This operating rope is attached to the shipper wheel on the machine and is revolved in a direction corresponding to that in which the hand rope is moved. Attached to the shipper wheel or to the shaft on which the shipper wheel may be mounted is a connection to the motor controller. According to the direction in which the shipper wheel is moved, the reverse switch on the controller is closed to a position to give the proper direction to the motor. The brake may be either mechanically or electrically operated. With the machines shown in Figs. 1-2 and 1-3 the brake *F* is mechanically operated. A cam *H* (Fig. 1-3), mounted on the inside of the shipper wheel, presses against a roller *R* in the lower end of bell crank *D*, which

Fig. 1-3.   Belted type of elevator machine for floor installation.

is attached to the brake lever at *E*. As the lower end of the bell crank is moved to the left, the end attached to the brake is raised, and with it the brake lever and weight *W*. This movement expands the brake band *F*, releasing the brake wheel, and leaves the machine free to be operated by the motor.

Automatic Stopping at Terminal Landings. At the top and bottom landings the machine is stopped automatically. Where a mechanically operated brake is used, as shown in Figs. 1-2 and 1-3, and the operating rope runs through the car, automatic stopping at the top and bottom landings may be accomplished in two ways. The first method is to place stop balls on the operating rope at locations where they will be hit by the car as it approaches the terminal landings; the stop balls then move the operating mechanism to the central position in much the same way as the operator would. The other method of stopping the car automatically is by the drum-shaft limits. These limits consist of a traveling nut T and two stationary nuts N and N', shown on the end of the drum shaft, which is threaded (Figs. 1-2 and 1-3). The traveling nut T engages the shipper wheel through a casting M, which has a tongue on its inner surface that fits into a groove in the end of T. On each side of T is a lip that, when the car reaches the top or bottom landing, engages a similar lip on the inside of nut N or N'. With this arrangement the winding drum L must turn in an opposite direction to that in which the shipper wheel is turned to start the car.

Fig. 1-4.   Drum type of elevator machine with motor mounted on same bedplate.

For example, if the shipper wheel is turned in a clockwise direction the drum must turn in a counterclockwise direction for proper operation of the elevator machine. In doing this with a right-hand thread on the end of the drum shaft, traveling nut T will move toward stationary nut N, which, if properly adjusted, will engage T when the car approaches the bottom landing. When this occurs, the shipper wheel will turn in a counterclockwise direction and come to a neutral position to stop the elevator by applying the brake. If nut N is in the proper position, the car will stop level with the bottom landing. When the car comes to the top landing, a similar operation takes place when T engages N'. In this case, however, the drum is turning in a clockwise direction and the shipper wheel is turned in a counterclockwise direction to impart this direction of motion to the car. Therefore, when T and N' engage, the shipper wheel is turned in a clockwise direction and brought to the neutral position.

Adjustment of the limits is made by adjusting the stop balls on the operating rope so that the car stops at the top and bottom floors with full load in car and brake properly adjusted. After this has been done, nuts N and N' on the drum shaft are adjusted so that they stop the car in practically the same position as the stop balls. Drum-shaft limits should be set slightly behind the stop balls on the operating rope, so that if these balls slip or the operating rope breaks or becomes fouled in the sheaves, the drum-shaft limits will stop the car at the terminal landings.

When the machine is operated from a handwheel or a lever in the car, the only limits to stop the car at the terminal landings are those on the drum shaft. On machines with an electrically operated brake (Fig. 1-4) a third method of limiting the travel of the car at the terminal landings is by opening the line switch and cutting the power out of the machine entirely. This is done by placing one switch at the top and another at the bottom of the hoistway so that they will be opened by the car if it goes by the terminal landings. Opening either one of these switches interrupts the circuit of the main-line switch holding coil, causing it to open and cut power out of the motor. Opening this switch also opens the brake-magnet-coil circuit, which applies the brake, stopping the machine. With a mechanical brake it would be of no use to cut power out of the motor unless the brake were applied, as the machine would probably keep in motion and control of it might be lost, with serious consequences. Some mechanical-brake designs incorporate an electromagnet which, when deenergized, permits the brake to apply on power failure (Fig. 4-5). See Chap. 4 for magnet-operated brakes.

Operation of Brake Mechanisms. With the Maintenance Company's machine (Fig. 1-4) the brake is released by magnet M pulling in two cores attached to an extension of the brake shoes at A. The brake shoes are fulcrumed on pins P and are applied to the brake wheel by springs W. This machine has a shipper wheel S on the end of the drum shaft to which the operating rope running to the car is attached, as in Figs. 1-2 and 1-3. It is also equipped with the drum-shaft limits T, N, and N'. The controller for the motor is attached to the shipper wheel by a chain that runs on sprocket wheel J.

There is one other safety device on the semimagnetically controlled elevator machine, that is, the slack-rope limit. When the elevator machine is equipped with a mechanical brake, as in Fig.

1-2, this safety device is mechanical in its operation.   The car-hoisting ropes come off winding drum *L* and pass up the hoistway against the small sheaves *I*.   While the ropes are taut, framework *Q* is held in the position shown.   Should the ropes become slack, because of the car safeties setting on the guide rails or the car striking the bumpers in the bottom of the hoistway, the framework drops and throws in a clutch on the back of the shipper wheel.   This clutch connects the drum shaft to the shipper wheel; the motion of the drum brings the shipper wheel to the off position in the same manner as the drum-shaft limits do at the terminal landings.   When

Fig. 1-5.   Worm and internal spur-geared drum-type elevator machine for slow car speed.

the machine has an electrically operated brake, the slack-rope device generally consists of a bar that runs across the bottom of the drum. This bar is attached to a switch, known as the slack-rope switch, in series with the holding coil on the main-line magnet switch.   If the ropes become slack, they will strike the bar and open the slack-rope switch, which in turn opens the main-line switch and causes the elevator machine to stop.

On the machines shown in Figs. 1-2 and 1-3 their winding drum *L* is driven by a worm and gear, the drum being keyed or bolted directly to the gear wheel and shaft.   With the machine shown in Fig. 1-5 a spur gear is keyed to the gear shaft, which meshes to an annular gear on the winding drum, thus giving two speed reductions from that of the motor. This allows the drum to be driven at slow speed by a high-speed motor.

In the machine shown in Fig. 1-5 the shipper wheel S and limits are not mounted directly on an extension of the worm-gear shaft, but on a short shaft mounted at right angles to it. These two shafts are connected by beveled gears U. On the end of the operating shaft is an eccentric E which releases and applies the brake through rod W. Part of the slack-rope device is under the drum and arranged so that, should the ropes become slack, clutch R is thrown into mesh with the shipper wheel through rod I, causing the machine to stop.

Elevator-machine Types. Electric elevators have been developed in many different designs, but in general they may be grouped in two classes: drum and traction types. On the drum type, one end of each hoist rope fastens to the car and the other end to the winding drum on which the ropes wind or unwind to hoist or lower the car. Even though thousands of these machines are in use in older-type buildings, they are not now permitted by The American Standard Safety Code for Elevators except for freight service where the lift does not exceed 40 ft and the car speed is 50 fpm or less. They can also be used on sidewalk elevators.

On traction-type elevators one end of each hoist rope connects to the car and passes over a grooved traction sheave to the counterweight, where the other end connects. The traction sheave is either mounted directly on the motor shaft or driven through a worm and gear. Friction between the ropes and sheave grooves transmits power from the driving motor to the car and counterweights. This design is suited to the highest lifts and car speeds and is generally used today for most electric-elevator applications except for some small-capacity low-rise slow-speed installations.

Classification of Control Equipment. The controls for elevator motors operated on direct current may be divided into:

1. Rheostatic, constant voltage, where resistors are used to reduce the starting voltage and perform other control functions.

2. Multivoltage (now obsolete), where means are provided to apply the starting voltage in definite steps, such as 25, 50, 75, and 100 per cent, as the motor comes up to speed.

3. Variable voltage, where voltage at the elevator driving motor terminals is gradually increased or decreased during the starting or stopping periods by adjusting the voltage of a generator supplying the motor.

The controls for a-c motors may be divided into two general classes: those for single-speed motors and those used with two-speed motors.

Examples of how these different controls operate will be found in Chaps. 5 to 9.

Basement Installation of Drum Machines. Drum machines have in general been built for two classes of installations—basement and overhead. Figure 1-6 shows a complete installation of an Otis basement-type machine for passenger service operated from a switch in the car. Two sets of ropes run from drum D to the car and counterweights. The car ropes run from the back of the drum up the hoistway over a sheave S and to the car crosshead C. The counterweight ropes come from the front of the drum and go under traveling sheave V, up the hoistway over the overhead sheaves S', and down to the drum counterweights W. A third set of ropes runs from the car to the car counterweights W'.

The ropes are arranged on the drum so that as one set winds, the other unwinds; this allows both sets to make use of the same grooves. To allow the ropes to pass to the drum counterweights, the car counterweights are slotted in back; the ropes are covered with steel tubes to prevent abrasion. The car counterweight is always above the drum counterweight so that, if the drum counterweight ropes break, this weight cannot fall on the car counterweight and overbalance the car. Otherwise the counterweights might fall, lifting the car and causing a serious wreck.

In Fig. 1-7 a chain H can be seen running from the bottom of the car to the bottom of the counterweights. This chain is used to compensate for the weight of the ropes being shifted to the counterweights when the car is at the top landing and shifted to the car when it is at the bottom landing. In Figs. 1-6 and 1-7 the car is near the top landing and the counterweights are at the bottom. In this case the weight of four ropes for the full length of the hoistway has been shifted to the counterweights and taken off the weight of the car. If the ropes weigh 400 lb, then 400 lb has been taken off the car and put on the counterweights, or there has been a change in weight of 800 lb between the car and the counterweight in traveling from the bottom to the top landing. In going from the bottom to the top landing, the weight of the compensating chain has been transferred from the counterweights to the car. Therefore, if the chain weighs the same as the ropes, as much weight has been added to the car as was removed and the counterweighting remains the same for any position of the car. A compensating chain or rope is

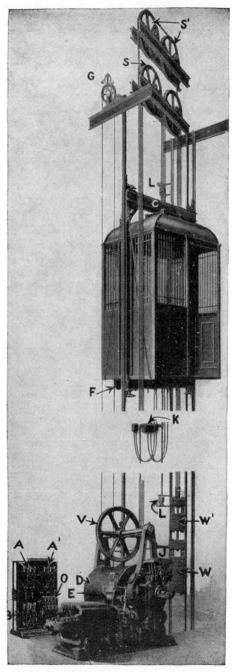

Fɪɢ. 1-6.   Typical basement installation of a drum-type elevator machine.

not necessary except on medium- and high-rise machines. Where chains are used, a hemp rope is laced through the links to prevent them from being noisy.

Overhead Installations of Drum Machines. On drum machines installed overhead (Fig. 1-7), the arrangement of the ropes is similar to that for machines installed in the basement, the chief difference being that there are no overhead sheaves. On the overhead machine the drum counterweight ropes come up the front of drum *D* and go down its back to the counterweights, where the car ropes come from the back of the drum and down the front to the car. The car-counterweight ropes come from the car up over traveling sheave *V* and down to the counterweights. This is the arrangement in which the drum diameter equals one-half the width of the hoistway. When the hoistway is wider than twice the diameter of the drum, deflecting sheaves are used for the counterweight ropes to divert them down the hoistway.

Arrangement of governor *G* and its rope can be clearly seen in Figs. 1-6 and 1-7. This rope runs over a sheave on the governor and around a weighted sheave *N* in the hoistway pit (Fig. 1-7). In Fig. 1-6 the two ends of the rope connect to a fitting *F* at the bottom of the car, and at the top, as shown in Fig. 1-7. This fitting is held in

FIG. 1-7. Typical overhead installation of a drum-type elevator machine.

a clip so that the governor rope moves with the car to drive the governor, equipped with flyballs similar to those found on a steam-engine governor. When the car runs above a predetermined speed, the governor flyballs are thrown out to trip two jaws that grip the rope and pull it out of clip *F* on the car. Pulling the governor rope free from the car sets the safeties under the car to prevent it from falling in case it gets out of control or the hoisting ropes break. For further details about governors and car safeties, see Chap. 3.

Controller Operation. The controller used on the machines shown in Figs. 1-6 and 1-7 is representative of those used on car-switch-controlled medium-speed drum-type machines, in that it has a main-line or potential switch *B;* two direction switches *A* and *A'*, one of which closes for up motion of the car and the other for down motion; and an accelerating magnet switch *O* for cutting out the starting resistance. When the operating switch in the car is thrown to the up position, one of the direction switches closes and completes the circuit through the motor's armature in one direction; when the operating switch is thrown to the opposite position, the other direction switch closes and completes an opposite circuit through the motor's armature, thus reversing its direction and that of the car. These direction switches are so interlocked that one must be open when the other is closed. When the direction switches close, they also close the circuit to brake magnet coil *E*, which releases the brake and leaves the elevator machine free to be operated by the motor.

To connect the operating switch in the car to the controller, a flexible cable runs from the car to a junction box *K* halfway up the hoistway. This gives a flexible connection that allows the car to travel the full length of the hoistway. From this junction box the wires generally lead to the controller in metal conduit.

On the machines shown in Figs. 1-6 and 1-7 there are two sets of limits. One of these at *J* consists of a number of contactors opened and closed by cams operated from a traveling nut on the drum-shaft end. The second set consists of the hoistway-limit switches located at *L* near the top and bottom landings. One set of contactors located at *J* on the machine is the first to open, as the car approaches the terminal landings, and interrupts the circuit to the direction switch, causing it to open; the car is then stopped by applying the brake. If for any reason the direction switch does not open and the car continues in motion, a second set of contacts will open shortly after the first and open the coil circuit to the potential switch *B*, causing it to drop out. This will cut the power out of the motor and apply the brake. Should the drum-shaft limits be out of adjustment and allow the car to go by the terminal landing, then a cam on the car would strike one of the hoistway-limit switches L and open it. Opening the hoistway limit breaks the potential-switch coil circuit with the same effect as if it were opened from the drum-shaft limits.

In addition to the limit switches mentioned in the foregoing, there is the slack-rope switch and, in some cases, a switch on the governor. As previously stated, the slack-rope switch is under the drum and is

Fig. 1-8. Worm and gear with ball thrust bearings at *T*.

in series with the holding coil on the potential switch. When the ropes become slack on the drum, owing to the safeties setting on the car or other causes, the slack-rope switch is opened. This in turn opens the potential switch and stops the motor. When a governor switch is used, in case overspeed causes the governor to operate and set the safeties, the governor switch opens and the motor is stopped by opening the potential switch.

Thrust Bearings. Another important part of a geared elevator is the thrust bearings. Because of the unbalanced weight of the car or counterweights, the worm gear exerts a thrust on the worm that must be taken care of by thrust bearings. One common arrangement is shown in Fig. 1-8, where ball-type thrust bearings are shown at *T*. In other cases roller thrust bearings are used; again, on older elevator machines, bronze and steel disks are employed. On some types of elevators both thrust bearings are included in the rear worm-shaft bearing. This latter arrangement places both thrust bearings where they can be inspected and adjusted without taking out the worm shaft, as had to be done on many of the older types of machines.

Instead of a single worm-gear drive (Figs. 1-6 and 1-7), a tandem worm and gear is used (Fig. 1-9). The rear gear is mounted on the drum shaft and meshes with the front wheel, the worm wheels being

FIG. 1-9. Tandem worm-geared drum-type elevator machine does not require thrust bearings on its worm shaft.

driven by tandem worms. This gearing automatically absorbs the thrust, and no thrust bearings are required.

Different makes of elevator machines will have different arrangements and construction of mechanical details, but all drum-type machines are essentially the same as those described in the foregoing.

Traction Machines.   Although the principle of the traction drive is old, its successful commercial application to an elevator has taken place within the last fifty years. With its drum limited as to length and diameter, use of the winding-drum machine is restricted to lifts of about 150 ft or less. The advent of the modern skyscraper building made it imperative to develop an elevator machine suitable for high rises, and the traction type came into commercial use. This machine has proved so successful that it has rapidly superseded winding-drum types.

1-to-1 Roping. Figure 1-10 shows a complete installation of an Otis double-wrap direct-traction machine, roped for a speed ratio of 1 to 1. Here the ropes pass from the car's crosshead at C, over the traction sheave S, around the secondary sheave S', then over the traction sheave S, and to the counterweight crosshead at W. Six ropes are usually used on these machines and, since they pass over the traction sheave twice, this sheave has 12 grooves. Because the ropes make two half wraps around the traction sheave, this type of equipment is known as a full-wrap or double-wrap machine. A machine similar to that shown in Fig. 1-10 is shown in Fig. 1-12, from which the way that the ropes pass around the sheaves may be more clearly seen.

Car speed depends on traction-sheave size and motor speed. Traction-sheave diameter must be large enough so that bending stresses in the ropes will not cause short rope life. Minimum traction-sheave diameter is about 27 in., or about 40 rope diameters, as limited by The American Standard Safety Code for Elevators. Maximum diameter is limited by car speed required and it must not exceed the span between the car and counterweight centers. A sheave 36 in. in diameter and a car speed of 600 fpm will require a motor speed of only $600/[(36 \times 3.1416)/12] = 63.6$ rpm; therefore, motor speed on a direct-traction machine is necessarily low. Because of these limitations the direct-traction 1-to-1 machine is limited to a car speed of over 400 fpm. In Fig. 1-10, car speed equals rope speed, and has what is called a 1-to-1 roping.

2-to-1 Roping. In some installations a roping is used that gives a ratio of rope speed to car speed of 2-to-1. With the exception of the method of roping up the machine to the car, the 2-to-1 direct-traction elevator (Fig. 1-11) is practically the same as the 1-to-1 machine (Fig. 1-10). In the 2-to-1 machine, one end of the hoist ropes dead-ends at D in the overhead beams and goes down around sheave C in the car's crosshead, then up around the traction and secondary sheaves S and S', and down around a sheave W in the counterweight crosshead, up to a dead end D' in the overhead. With this arrangement car speed is only one-half rope speed or one-half that for 1-to-1 roping with the same motor speed and same size traction sheave in both installations. The 2-to-1 direct-traction installation is generally limited to car speeds of 350 to 500 fpm.

Instead of the secondary sheave S' (Fig. 1-11) being directly under the traction sheave S as in Fig. 1-10, it is offset toward the

FIG. 1-10. Traction-elevator installation with 1-to-1 roping.

FIG. 1-11. Traction-elevator installation with 2-to-1 roping.

counterweight side of the hoistway (see also Fig. 1-12). This is done when the distance between the counterlines of the car and counterweight rails is greater than the diameter of the traction sheave. When this condition exists, the secondary sheave serves also as a deflector to lead the ropes vertically down to the counterweights.

Geared Traction Machines. In the geared traction machine a high-speed motor is used and the desired car speed is obtained by using the proper gear reduction. Motor speeds for this type of installation

Fɪɢ. 1-12. Gearless traction machine with ropes on traction and idler sheaves.

vary from about 400 to 1,200 rpm, depending upon the motor size and type of installation. The geared traction is used with either a 1-to-1 or 2-to-1 roping and is generally applied to car speeds ranging from about 50 to 350 fpm, with a few going to 700 fpm.

One of the features of a traction-elevator machine is that if either car or counterweight bottoms, traction is lost between the ropes and the traction sheave and the car or the counterweight will not be pulled into the overhead work as might happen with a drum-type machine.

In Fig. 1-10, in addition to the ropes running from the top of the car over the traction sheave to the counterweight, a second set of ropes runs from the bottom of the car around a sheave *T* in the pit to the lower end of the counterweight. These are the compensating ropes and compensate for transferring the hoisting ropes' weight from the car to the counterweight or vice versa, as explained for Fig. 1-7. Sheave *T* and its housing on the compensating ropes are free to move vertically on guides in the pit, and the weight of the sheave and housing is depended upon to keep a suitable tension in the compensating ropes.

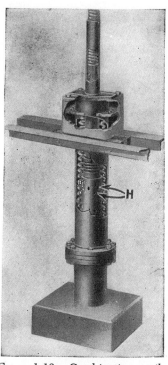

The compensating sheave provides certain safety features in that if either the car or counterweight bottoms, further motion lifts the sheave and tends to prevent the car or counterweight from being pulled into the overhead work, as might happen with very high rises. A switch is generally provided on the compensating sheaves so that if it lifts above a certain point the switch opens, this in turn opens the main-line switch on the controller and cuts the power out of the machine.

Oil-spring Buffers. Connected to the bottom of the counterweights is an oil-spring buffer (Fig. 1-10). A similar type of buffer *O* is in the bottom of the pit to arrest the motion of the car should it go below the bottom landing. This buffer is so constructed that it brings the loaded car to rest from full speed without serious discomfort to the passengers. A sectional view of the buffer is shown in Fig. 1-13.

Fig.   1-13.   Combination   oil-spring buffer, which is placed at bottom of hoistway to bring car to rest in case of overtravel at bottom landing.

As the buffer is compressed, oil is forced from the inner to the outer cylinder through holes indicated at *H*. When the piston is unloaded, the spring returns it to its normal position. It is now general practice to put both the car and counterweight buffers in the pit.

On high-speed traction machines safeties may be used on both the car and counterweight. In Fig. 1-10 one side of the governor rope passes over two small sheaves at *P*, then down around the drum on the counterweight safeties, and out over two other pulleys *P'*. From the counterweight the governor rope passes around tension sheaves *B* in the pit and then to the car, where it is deflected over small idler sheaves under the car and makes two turns around the drum on the car safeties.   From the car-safety drum the rope passes over an idler sheave to a releasing carrier at the top of the car and back over the governor sheave *G* to the counterweight.   Should the hoisting ropes part, the downward motion of the car and counterweight

would cause the governor rope to rotate the safety drums and apply the clamping jaws to the guide rails, bringing both car and counterweight to rest.

The brake wheel and traction sheave on some direct-traction machines are pressed and keyed on the armature shaft; on others, they are bolted to the armature spiders.   Since there is no reduction gear between the traction sheave and the brake, the latter must be strong enough to hold the maximum difference in weight between the fully loaded car and the counterweights. For this reason the brake is much larger than on a gear-type machine. The brake shoes are released by magnet *M* and applied by heavy springs *E* (Figs. 1-10 and 1-12).   When direction-traction machines are to be used for safe lifting or other heavy purposes, they are sometimes equipped with two brakes.

FIG. 1-14.   High-speed traction-elevator-machine car switch with cover removed.   The wheel is for operating the car safeties in an overspeed emergency.

Car-switch Control.   The control of the machine (Figs. 1-10 and 1-12) may be from a switch in the car, as for a drum-type machine. However, where the machine is used for medium speeds, it is possible to obtain as high as seven different speeds from the car switch. This is necessary not only for smooth acceleration, but also to allow the operator to make landing stops with ease.   Figure 1-14 shows a car switch with the cover removed for a traction elevator machine. The handwheel just below the switch, used on old-type machines, is for applying the safeties in case the car gets out of control.   Car-switch controls are now practically obsolete, being replaced by push-button types.

On account of rope creep on the traction sheave, machine limits to stop the car at the terminal landings cannot be used as for the drum-type machine. Two general schemes of limit stops are used. With one arrangement a series of switches is located in the hoistway at the terminal landings and a cam, which opens these switches, is attached to the car. The first switch to open slows the car down and the last brings it to rest.

In Fig. 1-10 the limit switches are mounted on top of the car at *A* and a cam *D* is mounted at both the top and bottom landings. As the car approaches the landing, the cam opens the limit switches and gradually brings the car to rest. Final limit switches *H* are mounted in the hoistway, so that if the car passes the landings it will open one of these switches, which in turn will interrupt the circuit to the potential switch on the controller and cut the power out of the machine.

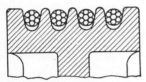 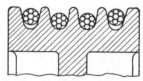

FIG. 1-15. Section of V-grooved sheave with all ropes and grooves the same size.

FIG. 1-16. Condition that can develop on a V-grooved sheave when ropes and grooves wear unevenly.

V-grooved Traction Machines. Another type of traction machine is that known as the V-grooved, sometimes called pinch-groove, half-wrap, or single-wrap machine. In such machines the grooves in the traction sheave are V-shaped and the ropes pass directly from the car over this sheave to the counterweights. No idler sheave is used, and the traction between ropes and sheave is obtained largely by the wedging effect of the ropes in the grooves. The ropes make one-half wrap on the traction sheave; thus the name half-wrap or single-wrap machine. Other types of traction machines have been developed and are in use, but those described give a good general idea of this class of equipment.

In the single-wrap machine, the maximum contact arc between the ropes and sheave cannot exceed 180°, and is less than this for equipments using a deflecting sheave. In modern single-wrap machines different types of grooves are used, but they all are shaped so that the ropes cannot bottom in the grooves, which allows a pinching action between the sides of the grooves and ropes (Fig. 1-15). One of the difficulties with this arrangement has been to make all the ropes travel at the same speed. To obtain this condition it is necessary for all sheave grooves and ropes to be alike. If one rope or groove wears more than another, as they sometimes do, a condition similar to that shown in Fig. 1-16 exists on the sheave and causes the ropes to travel at different speeds. They then have to slip in the grooves to compensate the difference in groove diameter. For further consideration of this problem see Chap. 13. It is now general practice to use undercut U's or preformed grooves (see A, Fig. 2-11) that maintain traction regardless of groove wear for long periods.

Stopping Car Level with Floors. Accuracy of landing and maintenance of the car platform level with the landing floor during loading and unloading are two important factors in elevator service. In passenger-elevator service, as car speed increases, landing the car accurately at the floors becomes more difficult. With the usual manual types of control greater dependence must be placed on the operator's skill to make the landing without undue slowing down of the service, either by slowing down the car too early before stopping or, because of the operator's failure to make the landing accurately, by having to inch the car either up or down.

Accurate landing is a factor in eliminating hazards from passengers' tripping. Also, when the car is stopped level with the floor, passengers move on and off the car in less time than when they must step up or down at the bidding of the operator. Therefore, accurate leveling of the car's platform with the landing floors will tend to improve the elevator service as well as decrease the time required to make a round trip.

If accurate landings can be assured, not only is it possible to render better and more service with a given number of elevators, but when elevators are being considered for a new building, it may be possible to reduce the number installed. This would result in a reduction in the cost of the elevator installation and increase the renting space in the building, an even larger factor.

In freight-elevator service stopping of the car level with the landing floors and maintaining this level greatly facilitates the movement of freight on and off the car. It also reduces the wear on freight-handling equipment, such as trucks, as well as the damage to merchandise being loaded on and off the elevator. Maintaining the car level with the landing floor will result in greater safety to employees handling the freight. One of the difficulties in keeping a freight elevator level with the floor is in taking care of the stretch in the ropes with a change in the load.

When heavily loaded trucks are being put on the car, if the latter's platform is maintained level with the landing floor, the elevator equipment will be relieved of the severe strains of the load dropping 2 to 3 in. or more on the car floor. Preventing these heavy shocks to the equipment will assist materially in reducing wear on the elevator, which not only reduces maintenance costs, but results in better elevator service.

To start and accelerate the elevator consumes a considerable percentage of the power required to operate the machine. If accurate

Fig. 1-17.   Microdrive applied to a direct-traction elevator machine.

landings can be made without inching the machine, the power consumption will be kept at a minimum. The power that can be saved by eliminating inching of the elevator will vary with the type of control, being a minimum with an adjustable voltage control and a maximum with rheostatic control. Prevention of inching also reduces the wear on the controller, ropes, and other parts of the equipment.

Microdrive Traction Machines.   A number of developments in elevator design have been made to level the car at the landings automatically, independently of the operator. One of the earlier developments (Fig. 1-17) is the Otis microdrive machine. Main motor M and traction sheave S are the same as for the standard direct-traction machine. In addition to the main machine a micromachine, with a motor M', brake B, and worm gear G, is attached to the main machine through suitable gearing. A small spur gear at *A* on the micromachine's worm-gear shaft meshes into a large gear in the case *A'*. The brake shoes are supported on the inside of this gear wheel, and the brake wheel is keyed to the traction-sheave shaft. The brake wheel and brake are so arranged that when the brake is released the main machine is free to move, as in the usual design. When the brake is applied, the micromachine is connected to the main machine. In other words, the brake and brake wheel on the main machine act as a clutch between this machine and the micromachine.

The micromachine is of small capacity and operates the car at a speed suited for making accurate landings. For medium-speed machines the micromachine gives a car speed of 30 to 60 fpm; on high-speed installations the micromachine car speed is 45 to 90 fpm. However, this does not slow down the elevator operation, since the microzone does not come into effect until the car is within about 10 in. above or below the floor for car speeds up to 400 fpm. For car speeds of over 400 fpm the microzone is reached when the car comes within 16 to 18 in. of the floor.

Microdrive Machine Controllers. With the older type of micro-leveling machines and car-switch control there are two controllers. One operates the main machine from the car switch. The other controller is for the micromachine and is brought into operation automatically as the car approaches the floor at which a stop is to be made by a switch mounted on top of the car and cams in the hoistway. The switch has two sets of contacts, one for up motion and the other for down. There are two cams at each landing. One operates the micromachine and brings the car up level with the floor and the other brings the car down level with the floor.

At the top and bottom landings the car is stopped and leveled into the floor automatically. At intermediate landings the operator must, on approaching a floor where a stop is to be made, bring the car under control, and when within a few feet of the landing bring the car switch to center, in much the same way as for the standard machine. Then when the car comes within 16 or 18 in. of the floor, the micromachine is switched in automatically and the car brought level with the landing and stopped.

The accuracy with which the micromachine will maintain the car level with the floor will depend upon conditions. For passenger service where the micromachine speed should be comparatively high, a 1/2-in. variation in the levels of the car at landing floors would be allowed. In freight service, where much slower car speeds are used, the variation would not be over ¼ in., while in general the car would be maintained absolutely level with the landing floor. With modern Otis elevator machines and those of other companies with push-button automatic control, leveling is done by the main machine, and the leveling machine is not used.

Electron-tube Leveling. Another method for automatically leveling an elevator car at the landings, developed by the General Electric Company, uses electron tubes. This system is applied with variable-voltage control, and leveling is done with the main machine.

These tubes have a filament, a grid, and a plate, as indicated in Fig. 1-18. They will oscillate if coils are arranged in the grid and

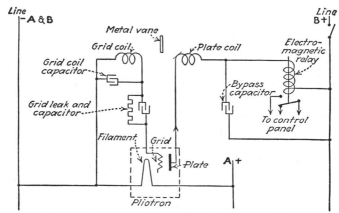

F<small>IG</small>. 1-18.   Diagram of an electron-tube elevator-leveling circuit.

the plate circuits in proximity to each other so that their fields couple, and if the grid coil is suitably tuned with a capacitor across it. The plate coil, grid coil, and grid-coil capacitor are shown in Fig. 1-18. The frequency at which the circuit oscillates is determined by the frequency of the tuned grid circuit. In standard General Electric elevator-leveling units this frequency is about 200 kc.

Inasmuch as the coupling between grid and plate coil is essential for oscillation, breaking this coupling will stop the oscillation. Figure 1-19 shows one of the car-leveling units. *P* is the electron tube; the way the grid and plate coils are mounted back of the unit is indicated at *C*. How these coils are mounted is more clearly shown in Fig. 1-20. Inserting a metal vane in the space between the two coils prevents coupling of their fields and stops the oscillation of the circuits.

In an oscillating circuit of the kind described a considerable change in plate current occurs when the circuit goes in or out of oscillation. The plate current will be low when the circuit is oscillating and high when oscillation ceases. If an electromagnetic relay is connected in series with the plate circuit and the relay coil bypassed with a capacitor, as in Fig. 1-18, the difference in plate current through the relay coil will be further increased between the oscillating and nonoscillating states, because the highly inductive

Fig. 1-19. Electron-tube car-leveling unit with the cover removed.

Fig. 1-20. Coils mounted in *C* are in the plate and grid circuits.

relay coil will not pass radio frequencies, but the capacitor in multiple with this coil will.

In a nonoscillating condition the capacitor will pass no current and the relay will freely pass the direct current. The operation then will be as follows: with the vane between the grid and plate coils, the relay will be closed by the direct current; with the vane absent and the circuit free to oscillate, the relay will be open. The unit in Figs. 1-19 and 1-20 has approximately 1-in. air space between the grid and plate coils; $\frac{1}{16}$-in. radial movement of the edge of the vane between the coils is sufficient to pick up or drop out the relay. This pick-up and drop-out distance may be reduced if requirements demand.

Figure 1-21 shows a group of electron units mounted on the crosshead of an elevator car as used in a preregister signal-control elevator system. On the photo, R is a car guide rail, S a car guide shoe, H the car crosshead, T electron tubes, U a leveling unit attached to the car crosshead, and V one of the metal vanes clamped

Fig. 1-21. Electron-tube leveling units mounted on the car's crosshead. H is the crosshead; R, a guide rail; S, a guide shoe; T, electron tubes; U, one of the electron-tube units attached to the car's crosshead; and V, one of the metal vanes, clamped to a guide rail, that acts on the electron-tube circuits to level the car at the landings.

to a guide rail, which acts on the leveling-unit circuit to level the car at the landings.

This leveling system does not make a mechanical connection of any kind in the hoistway with the control equipment. Slowing down and leveling the car at the floors is accomplished by metal vanes, properly located in the hoistway, passing through the space between the grid and the plate coils on the electron units, which in turn cause relays to carry out the slowing-down and stopping operations.

Permanent-magnet Leveling. An automatic leveling device (Fig. 19-5) by Haughton Elevator and Moline Accessories uses the constant force of magnetic attraction to provide accurate leveling control of the car at the floors. The switch units, containing powerful permanent magnets and silver-contact assemblies, are mounted on the elevator car and actuated by proximity of steel vanes fastened to a guide rail at each floor. There is no mechanical contact between the switch units and the vanes. Accurate floor leveling is obtained even though the space between the switch unit and vane may vary because of guide-shoe wear on the use of roller guides.

Fig. 1-22. View of the top of the car and the rear of the hoistway.

Fig. 1-23. Stopping inductor switches.

Wiring circuits for these units are similar to those for other types of leveling. They may be used for any car speed and with d-c or a-c control. Each switch unit is totally enclosed in a nonmagnetic case for protection against dust and particles of metal which might be attracted by the permanent magnets. Mountings of these units and vanes are made adjustable to permit accurate setting of the leveling points.

Automatic-landing Equipment. Methods for automatically stopping elevator cars level with the landings have been developed. To make the stops accurately it is necessary that stopping begin at the right distance from each floor. Automatic landing was made possible by the development of variable-voltage control, which allows very accurate control of the car's speed. The highest-speed elevators in the world, operating at 1,400 fpm, in the Rockefeller Center main building, New York, are automatic-landing type.

A control for automatic-landing elevators, known as the automatic-inductor type of control, has been developed by the Westinghouse Electric Elevator Company. With the inductor control all operations for automatically slowing down the car and stopping it accurately at landings are accomplished without any mechanical contact between control parts on the car and stationary parts in the hoistway.

In the hoistway, mounted on the counterweight guide rails, are what are known as inductor plates, such as are shown at A and A' in Fig. 1-22. On top of the car are slow-down inductor switches B and stopping inductor switches C. These switches are totally enclosed in metal cases to protect them from dust and dirt. On the side of the cases next to the inductor plates nonmagnetic material is used.

The inductor switch consists of a coil and magnet and a normally closed contact (Fig. 1-24). As shown in Fig. 1-24, there are two inductor switches, the first slow-down switch and the second slow-down switch. With car-switch control, when the operator centers the car switch, coils C become energized and magnets M excited. When a magnet comes in front of its inductor plate, it is attracted and opens its contacts. Opening the contacts interrupts the circuit to the slow-down relay on the elevator-motor control board and decreases the car speed.

In Fig. 1-22 inductor plate A is out of line with A'. This allows No. 1 slowdown to function first, and as the car continues in motion the second slow-down inductor switch is operated by plate A'. A final stopping of the car is done by the stopping inductor, which contains three switches. This inductor switch is shown in Fig. 1-23 with the cover removed. The final slowdown and stopping are accomplished by an inductor plate, not shown in Fig. 1-22, which causes stopping inductor switch C to function as the car approaches the floor. The inductor plates in the hoistway are adjusted to give a minimum stopping distance and obtain a smooth and accurate stop.

Preregister Control. Elevators of the automatic-landing, or the automatic-leveling, car-switch type, traveling at high speed, require a more complete signal system than do ordinary car-switch-controlled machines. With the latter, waiting passengers on landings signal the car operator by pressing a button that indicates on an annunciator the floor number and direction in which the passenger wishes to go. On another system a flashlight in the car signals the operator, when approaching a floor, that passengers are waiting to go in the direction of car travel.

These systems function satisfactorily where cars have open-grille

Fig. 1-24. Inductor switch for first and second slowdowns.

gates and travel at moderate speeds. As car speed increases it becomes more difficult for the operator to read floor numbers and to judge stopping distances. These difficulties may cause a large number of false stops to be made or the car may be prematurely slowed down before making a stop, causing a reduction in service rendered by the elevator. Where cars with solid doors are used they must be stopped at the floors automatically, as the operator cannot see the landings. On car-switch-controlled cars the signal system must be so designed that it indicates to the operator when he should center the car switch, after which the car stops level with the floor automatically.

A preregistering signal system gives the operator a signal to center the car switch as the car enters a stopping zone at full speed. Auxiliary signals are given to the operator when the car is at the first of two successive floors at which stops are to be made. The system may also be used to slow down the elevator automatically from the registered stop signals if desired.

A preregister signal system that accomplishes these requirements

has been developed by the Elevator Supplies Company. With this system there are the usual up- and down-signal buttons at the landings. There are signal lanterns over each hoistway door to indicate to waiting passengers which cars are approaching the floors and the direction in which they are going. This arrangement is the same as is frequently used with elevators having the conventional type of car-switch control.

Above the car switch in each car is a panel on which there is a button for each floor, as shown in Fig. 1-25. The operator presses buttons corresponding to the floors called by passengers when entering the car. When a button is pressed it is held in the closed position, indicating to the operator where stops are to be made. The closed button also completes a circuit to a selector in the machine room. As the car travels up or down the hoistway a contact is made on the selector that flashes a signal light for the operator to center the car switch. This light is in the center at the top of the panel.

With this particular type of elevator control the signal light is an added precaution, since the operator does not have to depend on it to know when

Fig. 1-25. Car switch with pre-register-signal push buttons.

the car switch is to be centered. Slowdown is initiated by the registered signal, after which the operator centers the car switch. If the operator does not notice the car slowing down, the signal light's flashing shows that the car switch should be centered to make a stop.

The car- and hoistway-door control contacts are on the car switch. When the operator moves the switch to a start position, the doors close and the car starts and accelerates to full speed. With one type of control, when the car enters a stopping zone, as established by a button being pressed in the car or at a landing, the car slows down, after which the operator centers the car switch. The car then stops level with the floor for which the signal was given, and the car and landing doors open automatically.

When signals are registered on the car operator's panel, the buttons, being of the mechanically locked type, remain in closed position until the car approaches a terminal landing, when they are released automatically. Car-signal panels contain one button for each floor served, the same button being used for either direction of travel. If it is desired to reverse the car between terminal landings, the push buttons can be manually reset by button *R*. Should the operator be unable to take on additional waiting passengers, floor calls may be bypassed by switch *B*. Placing this switch in bypass position will transfer registered passenger signals to the next approaching car, but will not affect registered stop signals on the car operator's panel.

Automatic Dispatching. At the top of the panel the two small lights are for an automatic dispatching system. One light flashes when it is time for the operator to leave the first floor and the other flashes when the car should leave the top landing. The automatic dispatching system not only relieves the dispatcher of signaling the operators when to leave the first floor but it also ensures that they will be signaled at the proper time at both terminal landings. This assists the operators to run closer to schedule than they could by manual dispatching.

The small switches just above and below the car switch are part of the elevator control system and are for the lights and fan in the car. They are not a part of the signal system.

In the control room there are selector commutators, relay panels, and other equipment essential to the signal and control system. The commutators for the floor stops, floor lanterns, night annunciator, car-position indicator, and other operations are assembled in a single unit, as shown in Fig. 1-26. Vertical commutators of the straight-line double-screw type are used, except when the car speed is slow and the segment requirements are such that a standard friction-reversing commutator can be used. The commutator contacts brush is driven from the elevator's secondary sheave by a chain and sprockets. Stop nuts are assembled and locked on the screws and a friction stud is used on the drive shaft, so that commutator-brush position will be corrected at each end of travel if necessary.

Signal Operation. To take full advantage of the high-speed elevator automatic stopping became almost imperative, and it is to handle such service that signal operation systems have been developed. In these systems the operator only starts the car, after which it accelerates to full speed and goes to the first floor for which it was signaled, stopping level with the floor. The car and hoistway doors also open automatically as the car comes level with the floor, so that when the car stops, the doors are open. This is accomplished in such a manner that there is no danger to the passengers.

An Otis car-operator's control board is shown in Fig. 1-27. As

Fig. 1-26. Front of the commutator assembly for the floor stops, floor lanterns, night annunciator, car-position indicator, and other operations.

passengers come on the car and call the floor they want, the operator pushes a button corresponding to this floor. These buttons are shown at the top of the control board. Below these buttons is an operating lever *L*, similar to the ordinary car-control-switch lever. This lever has four positions—off, open, close, and start. The open and close positions of the lever are for the control of the car and hoistway doors. The start position initiates the starting of the car, after which the lever can be allowed to come to the off position. When the operator starts the car, the operating lever is moved to the start position and the landing and car doors close, after which the car starts automatically. The operator has no further action to take until the car stops at the first floor for which a button has been pushed. The button may be in the car or at a floor landing. After passengers have entered or left the car, the operator again pushes the control lever to the start position, when the hoistway and car doors close, and the car starts and goes to the next floor to which it has been signaled.

Until the car door closes and the landing doors close and lock, they are under the control of the operator. For example, if the control lever were in the start position and the doors were partly closed, and for some reason it was desired to open the doors, the operator could move the lever to the open position, as shown on the right of Fig. 1-27, and the doors would reverse and move to the open position. Closing the car and landing doors and locking the latter is what completes the control circuit to start the machine.

On each floor there are two buttons, one for up and the other for down direction. Each floor also has an up and a down signal lamp over the landing door of each elevator. Waiting passengers on a floor push either the up or the down button as they would on the ordinary type of elevator, but instead of signaling the operator to stop at the floor, the controller is set to stop the car automatically at the floor for which the button has been pressed. The first elevator approaching within 30 ft of the floor in the direction indicated by the signal will light the signal light and stop at the floor and the landing doors and car gate will open, all automatically. When the doors are again released by the operator, the car starts automatically and answers the next signal registered in the control system either by a waiting passenger or by the operator in the car. In all these operations the only functions performed by the operator are pushing buttons corresponding to the floors called by the passengers in the car and operating the lever that closes the doors and starts the car.

In the car there is only one set of buttons for both the up and the down directions. On the up direction, as the buttons for the different floors are pushed, they remain in the closed position until the car reaches the slow-down at the top landing. As the car is slowed down and stopped at the top landing, all buttons are released and the control placed in position for down motion.

Switches on Operator's Control Panel. Just above the operating lever there are three small switches—safety switch *A*, slow-speed *B*, and nonstop *C*. Moving the safety-switch button down opens the potential switch on the control board, cuts off the power from the machine, and applies a strong dynamic brake to slow the machine down, after which a mechanical brake is applied to stop the machine. This gives a positive stop by a separate means from the ordinary car-switch stop.

Pushing the slow-speed switch B to the down position cuts out part of the acceleration magnets on the control, so that the elevator cannot come up to full speed. This button is used when it is desired to operate the car at reduced speed.

If the car is full and the operator cannot take on any more passengers, pushing the nonstop button to the down position will bypass the floor signals, preventing the car from being stopped by a waiting passenger pushing a floor-signal button. The floor signals, however, are simply switched to the next car following; they are not removed from the control system. When a waiting passenger pushes a signal button on a floor, a car must come to this floor and stop before the signal is removed from the control system.

Below the operating lever are three more switches—micro D, emergency cutout E, and light F. The machine may be operated at slow speed by throwing the micro switch, which allows the elevator to travel at slow micro speed and, in case of an emergency due to a failure of the main motor or controller, allows the car to go to the landing. The emergency button, which is normally enclosed and cannot be pushed without breaking a glass cover, is used to bypass the gate contacts, so that the car can be moved with the gates open in an emergency. The light switch is for the lights in the car.

A reverse switch G is shown near the bottom of the control board. This is used chiefly when the car is being worked on to reverse the motion without going to the terminal landings. For example, if the car was started from the bottom landing and run to some intermediate floor and then it was desired to return without completing the trip, the reverse switch could be used to set the control buttons for down motion.

At the right bottom of the control board is a red light and lock switch H. These are for the motor-generator set used on what is known as the unit multivoltage control system. Putting the key in the lock and turning it in one direction starts the motor-generator set that powers the elevator motor. When it comes up to speed the red lamp lights, after which the key may be removed from the lock. Putting the key in the lock and turning it in the opposite direction shuts the motor generator down.

A large part of the equipment on the operator's control panel is duplicated on the dispatcher's board. On this board are also indi-

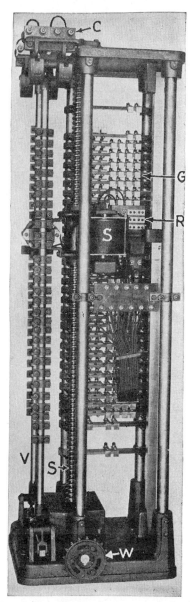

Fig. 1-27.   Car-operating control panel complete with push buttons and switches.

Fig. 1-28.   Floor selector on which all signals are registered for stopping the car at the different floors.

cated the position of each elevator, the direction in which it is going, the waiting-passenger signals indicated from the floor landing buttons, and whether the cars are running on a nonstop schedule or answering the floor calls. If the dispatcher finds that the cars are beginning to close up on one another, he may throw the nonstop switch on one and thereby allow this car to speed up and gain headway. If an operator switches to nonstop operation, this immediately shows on the dispatcher's board, so that if he wishes to investigate the cause he may be at the landing to meet the car when it arrives.

Microleveling Signal-control Machines. The chief difference between a signal-control machine and the standard machine with hand control is the addition of the signal-control equipment, one of the chief parts of which is the floor selector (Fig. 1-28). A steel tape runs from the top of the car over an overhead sheave, down around a tension sheave at the bottom of the hoistway, and to the bottom of the car. This tape forms a positive drive for the selector. On the opposite end of the shaft from the overhead sheave a sprocket wheel connects through a chain to another sprocket wheel W on the driving shaft of the selector (Fig. 1-28). In this way any movement of the car is transmitted to the selector and the relation between the two is not affected by any stretch or slipping of the hoist cables. Sprocket wheel W on the selector is geared to the vertical screw shaft S, which carries the selector switch R. The retardation switches for the floors are operated by two vertical shafts V, having a number of U-clamps attached to them, one for each floor. The vertical shafts are connected by bell cranks and a rod at the bottom, and each is mechanically connected to a contactor C at the top end.

When a button is pushed in the car or at one of the floor landings, one group of contacts, such as G in Fig. 1-28, is made alive. When the selector switch R has moved onto this group of contacts, the slowdown circuit is completed and coil S on the selector is deenergized, releasing a pawl that catches under a U-clamp corresponding to the floor at which the elevator car is to stop. Further movement of the selector switch carries the rods V up or down, opens the contactor C, and completes the process of cutting power out of the main machine. This brings the car within the microzone at the floor; the car is stopped level with the floor, and the doors and gates open automatically. When the doors and gates are again closed by the operator, coil S is energized on the selector and pulls the pawl away from the floor limits. These return to their normal position, allowing contacts C to close and the car to start and go to the next floor that has been signaled.

Automatic-landing Push-button Control. In the Rockefeller Center main building the elevators are Westinghouse automatic-landing type, and the high-rise cars operate at 1,400 fpm. Variable-voltage full-automatic control is used and landing is done by the main motor. An inductor selector having a positive drive and direct ratio between car travel and selector movement is used. The selector carriage functions only in the zone of the floors served, so that its over-all height is greatly reduced.

The high-rise cars have a travel of 779 ft. This introduced a problem at the ground floor, due to stretch of the ropes when loading and unloading the car. A stopping and leveling relay using a photoelectric tube brings the car back to the floor when it moves away ½ in. because of rope stretch. The car and hoistway doors are protected by a light beam directed across their openings onto a photoelectric tube. Interruption of the light beam automatically prevents the doors from striking or injuring an entering passenger.

In the cars, the operators' panels are comparatively simple, as shown in Fig. 1-29. There is the usual bank of floor-call buttons at the top of the panel. These buttons are of the lock-in type, but each can be released after being set by a small button below it. On each side of the bank of call buttons are two rows of numerals, one for up direction

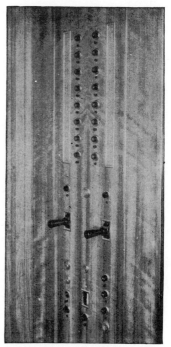

Fig. 1-29. Operator's panel in one of the highest-rise cars in RCA building, Radio City, New York.

and the other for down corresponding to the floor numbers. The numerals are illuminated by the waiting-passenger calls from the different floors. By this means the operator knows at all times on what floors passengers are waiting.

Just below the bank of floor-stop buttons are three others. The one on the right is for bypassing the floor calls; the center one is the emergency stop. If this button is pressed the car will stop as quickly as can be done safely. The button on the left is the next-stop control. If the car is in motion, and for any reason it is desired to stop, pressing the next-stop button will slow the car down and it will make a stop at the next landing reached by a normal stop. For example, a seven-floor run is required to make a normal stop from 1,400 fpm. If the car were operating at 1,400 fpm and the next-stop button were pressed, the car would stop at a floor seven floors away from where the next-stop button was pressed.

When the car has been stopped by the next-stop button, it may be started in either direction by pressing the terminal button for the direction desired and moving the control handle to the normal-run position. If the car is running in the up direction when the next-stop button is pressed, then after the car has come to rest, if the bottom terminal button is pressed, the car will start down and operate in a normal manner when the operating lever is moved to the run position. When the car is stopped at an intermediate landing, its direction may be reversed by pressing the button corresponding to the opposite terminal from that toward which it was traveling and moving the operating lever to the run position.

At about the center of the control panel are the two operating levers. The one on the right is for normal automatic control. Moving the lever up will close the doors, but the car will not start. This position of the operating lever is intended chiefly for checking operation of the doors and for closing the doors when the car is waiting its turn at the ground floor to take on passengers. When the control lever is moved to the down position, the doors close and the car operates normally under automatic control.

By pressing a button at the bottom of the panel, operation is changed from automatic to manual control by the left-hand operating lever. On manual control the car will run at a maximum speed of about 100 fpm.

Below the control levers is a key switch for starting and stopping the motor-generator set and a light that shows red when the set is running. There are three switches in the control circuit of the motor-generator set. One of these is on the starter's panel on the ground floor, one in the car, and one on the control panel in the machine room. All are connected in series. Consequently, if the motor generator is shut down from one location it cannot be started again until that switch is closed. This arrangement ensures against the machine being started unexpectedly when someone may be working on it.

At the bottom of the control panel in the car there are also two switches for control of the fan in the car, a switch for the car lights, a switch for the electronic door protection, a telephone jack for use when working on the car and wiring, a buzzer controlled from the starter's panel, a signal reset button, a car- and a hoistway-door by-pass under glass and, as previously mentioned, a button to switch from automatic to manual control.

Two of the high-rise cars are equipped for night service. These cars, in addition to having an operator's panel for express service, have an operating panel for night service. On the night-service panels are buttons and annunciators for all floors not served by express service. A combination control from the night-service and the express panels allows the elevator to serve all floors. The controls on these cars have both an express-service and a night-service selector. To change from the express- to the night-service selector the operator pushes a button on the night-service panel and the change is made automatically. When only the night-service cars are running, pressing a button on the starter panel illuminates signs on each floor that read "Use Night-service Cars."

Dispatcher-type Automatic Control. In large warehouses, the control of freight elevators has in some cases been made fully automatic from push buttons and the control placed in charge of a dispatcher. Those wanting an elevator signal the dispatcher, who pushes the button to bring a car that is not in service to the floor where it is required. When a car is loaded, the dispatcher is signaled and he sends it to the floor to which the load is to be taken. When the car is unloaded, closing the doors and gates transmits to the dispatcher the signal that the car is available for use. This system of control is known as the dispatcher-type automatic. Such machines are usually equipped with automatic floor-leveling devices to ensure easy loading and unloading of the car. This system of control, where it can be used, provides maximum use of the elevator equipment, as the machines are under the direction of a single station, where the available machines are indicated at all times.

Department-store Automatic Control. Department-store passenger elevators are generally stopped at each floor to discharge and take on passengers. For this service a system of automatic control has been developed, known as department-store control. The car gate and landing doors are power-operated and controlled by the operator from a small lever in the car. When the operator closes the gate and doors by moving the control lever to the closed position, the machine starts and takes the car to the next floor, where it stops and the gate and landing doors open automatically. The operator's function is to close the gate and landing doors by manipulating the control lever in the car. Otherwise the control is entirely automatic and the car stops at every floor.

Collective Control. A collective system of control has been developed for automatic elevators in apartment houses, hospitals, and similar buildings. This type differs from the conventional automatic type of push-button elevator control in that it will answer all calls automatically whether or not the car is in service when the landing buttons are pushed to signal the car. For example, assume that a passenger gets on the car at the bottom landing and pushes the button to go up. On other floors buttons may be pushed by passengers also wishing to go up. When the car reaches these floors it will stop automatically for the passengers, and after the doors and gates have been closed it will continue to the floors above for which buttons have been pushed. After the up calls have been answered the machine will then automatically answer the down calls that have been registered in the control system.

In some cases a second car is made to come into service and help take care of the calls when the car in operation has a full load. For instance, assume that a full load of passengers gets on the car at the top floor and that passengers at some of the lower floors have pushed buttons to be taken down. The car platform is so arranged that the load on it closes contact and the second car comes into service automatically and answers the down calls that the loaded car cannot take care of, the loaded car responding only to buttons pushed by the passengers in the car.

Control for these cars may be arranged for attendant operation as well as automatic, which is known as dual collective control. During periods of heavy travel, such as occurs in the mornings and early evenings, the elevators may be operated by an attendant. At other times the cars run full automatic, operated by the passengers. A change from automatic to attendant control is made in the car by setting a simple switch for the desired operation.

Corridor Control. Until recently a waiting corridor passenger pushed a button to call the elevator to the floor. When the car arrived and stopped and the passenger got on he or she pushed another button for the desired floor. Now a control system has been developed by K. M. White, which is known as corridor control, and combines the two calls into one. With this system all call buttons, one for each floor, are in a panel near the elevator entrance on each floor. Building patrons just push a button in this panel corresponding to the floor they want.

If a passenger is on the third floor and pushes a fifth-floor button, this tells the control that the call is for up direction. Then, the car on its up trip stops at the third floor to answer the call. After the passenger gets on and the doors close, the elevator starts automatically and, if a fourth-floor call is not registered, the car goes directly to the fifth floor and stops, and the doors open. After the doors have remained open for a preset period to let passengers off or on, the doors close and the elevator continues to answer up calls. When all up calls are answered the car starts down to answer down calls that have been registered by waiting passengers pressing corridor buttons on different floors.

After the passenger presses a corridor button for the floor to which he wishes to go, the elevator carries out his wishes without further effort on his part. This eliminates crowding around to push buttons in the car. A photoelectric control tells the elevator when the passengers are clear of the doors. Corridor control is applicable to one- or two-car operation. For a description of modern automatic controls for banks of passenger elevators, see Chap. 10.

Self-supporting or Underslung Elevators. All the elevators that have been considered are supported from the building structure and generally require a penthouse on the roof for the overhead sheaves of the elevator machine. Figure 1-30 shows an Otis light-freight elevator designed to be self-supporting on the cars, two guide rails *R*, and a steel column *C*. This design can be installed in a new or existing hoistway without building reinforcement, overhead supports, or a penthouse. It is intended to serve two- or three-story buildings with a maximum rise of 35 ft. and is available

Fɪɢ. 1-30. Light-freight self-supporting or underslung elevator.

in capacities of 1,500, 2,000, or 2,500 lb at a car speed of 25 fpm.

Guide rails and supporting columns are attached to the building at each floor and at the top of the hoistway to steady the structure and take care of any horizontal thrust caused by loading and unloading the car. Mounted as a complete unit on a concrete foundation near the bottom of the hoistway, the drum-type elevator machine is powered by an a-c motor. There are two speed reductions between the motor and hoisting drum. The first is a worm and gear and the second an external helical-cut pinion and gear.

A 2-to-1 roping from the machine drum to the car forms a third speed reduction. Two steel hoisting ropes dead-end overhead at D, run down around two sheaves under the car up over sheave S, and then down, connecting to the hoisting-machine winding drum. This method of roping doubles the lifting capacity of the machine, halves the tension in each rope, and helps to compensate for the lack of counterweights, which have been omitted to simplify the equipment.

The elevator is operated manually by constant-pressure push buttons at each landing and in the car. It has an overhead governor G, car safeties, terminal and final limit switches T and F, and a slack-rope switch. Electrical interlocks prevent the car from moving away from the landings unless the hoistway doors are closed and locked, and prevent these doors being opened unless the car is at a landing.

Other designs of self-supporting or underslung electric elevators are powered by a traction machine and have counterweights. They are built for either freight or passenger service for rated lifting capacities of 1,200 to 4,000 lb at car speeds of 75 to 100 fpm. Even though these elevators are comparatively simple in construction they comply with The American Standard Safety Code for Elevators.

Screw Lifts. Another form of self-supporting elevator, a Haughton screw lift, is used for general low-rise service where a penthouse is not practical and available space is small. It combines electric-geared elevator smoothness and positive control with the electro-hydraulic elevator's space-saving simplicity; see Chap. 18. The machine unit (Fig. 1-31), installed in the hoistway pit, occupies space largely unused by other types of elevators.

A nonrotating screw column C, fastened under the car in a rubber-cushioned torque absorber, is raised by rotating a bronze lifting nut carried on heavy oversize tapered roller bearings. This nut is driven by motor M through multiple-V belt B. A brake operating on a drum below the driven sheave is released electrically by electro-magnet E and applied mechanically by spring S for accurate, non-creep floor stops.

The screw goes down into a cylinder filled with oil in the ground, assuring constant lubrication. As the screw descends into the cylinder, the oil it displaces rises into a small surge tank. This type of elevator meets all requirements for safe operation. Safety features include normal and final limit switches, spring buffers, and a rubber-cushioned stop. If the car overtravels the top or bottom landing because the control switches are held in by hand it is stopped at the bottom by spring buffers and at the top by a pin and collar on the lower end of the screw, which engages the bottom side of the nut. When the pin and collar come against the nut, the stop is cushioned by springs on the underside of the car which allow a small rotation of the screw.

On the bottom end of the screw, a plate fitting the inside of the oil cylinder acts as a piston on the oil to hold the car speed at a safe value even if the nut is removed from the machine. On a drop test

FIG. 1-31. Screw-lift elevator machine used for general low-rise slow-speed service.

with the nut removed, down speed did not exceed 130 fpm, and the car was brought to a smooth stop by the spring buffers.

Maximum travel is in inverse ratio to the total load on the screw. A 20-ft rise is the limit for a total load of 13,000 lb (car plus rated capacity) on the screw. A total load of 8,500 lb could be lifted 25 ft, and 30 ft is the maximum for 4,900 lb. Higher travels require the use of a special screw.

Manlifts. In industrial plants, garages, and similar multistory buildings, where there are vertical processing operations, where frequent quick inspections of machinery on different levels are required, and where employees must be quickly dispatched to and from different floors as required by the work load, a type of elevator called a manlift may be installed. It may be considered as a highly refined development of a bucket elevator which, instead of having buckets to handle loose material, has steps on which a man can ride. As shown in Fig. 1-32, the Humphrey manlift has an endless flat belt 12, 14, or 16 in. wide that runs on a head and a foot pulley.

The foot pulley and its bearings are part of an assembly fastened to the bottom floor or to the floor in a pit. This pulley has a vertical screw adjustment to maintain correct tension in the belt. A gear-head motor *M* with integral disk brake and reduction gear coupled directly to the head pulley drives the belt and its load. V-belts are also used to connect the motor to the driving machine, as are right-angle gear drives.

Nonskid rubber-tread steps *S* on which a man can stand and ride are attached to

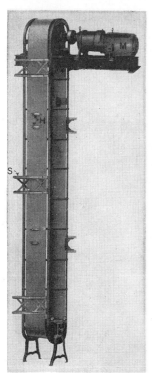

FIG. 1-32. A vertical man-lift has steps *S* on which men may ride from floor to floor.

FIG. 1-33. Four-roller step for a manlift-type elevator.

the belt at not less than 16 ft intervals. Four-roller steps are shown in Figs. 1-32 and 1-33, but two-roller steps may be used where the service is light. Flanged rollers on the steps running in either a 3-in. channel or angle track help guide the belt and support the steps in a horizontal position.

A light steel structure inside the belt from head to foot pulleys gives rigidity to the assembly. When the steps go over the head pulley they reverse so that a man can ride down on them as well as up. Between the steps are two hand holds *H*, one for up and one for down, which a man can grasp for support when riding on a step.

These lifts are equipped with safety devices to protect the riders fully, even against going over the head pulley. Movable safety cones are installed on the up-travel side of the lift just below all floor openings to discourage riders leaning out too far and being hit by the floor above. When a cone is bumped it opens a switch that immediately stops the lift.

A control rope running the full length of the lift on both sides provides a means of starting and stopping. During periods of light use the lift is normally stopped and is started only when used. During heavy traffic the lift runs continuously.

Limits to Uses. Maximum speed of either Humphrey or Ehrsam manlifts is limited to 80 fpm, and they are not generally installed for lifts above 200 ft. Even though they are completely equipped with safety devices, only plant employees familiar with their operation are permitted to ride on them, the public being barred. Package goods, construction materials, or tools are not permitted on manlifts, except those that fit entirely within the pockets of the usual working clothes of workmen.

An exception is made to this if tools are carried in a canvas bag, not larger than 11 by 13 in., provided with carrying loops or handles. Such a bag shall have a leather bottom but not shoulder straps and must be carried in the workman's hand while he is riding on a manlift. Conditions under which manlifts may be safely installed and operated are given in The American Standard Safety Code for Manlifts.

Freight Elevators. Passenger elevators and light-freight designs have been considered so far in this chapter. However, a large part of what has been said also applies to freight elevators. Their controllers and machines are practically the same, except that passenger elevators operate at higher car speeds, up to 1,400 fpm, with capacities up to 5,000 lb. Heavy freight is limited to speeds of about 450 fpm, and elevators can be built for any capacity—loaded freight cars or the largest trailer trucks. Where the speed of many passenger elevators permits using direct-traction machines with 1-to-1 roping, most new freight elevators have geared traction machines with 1-to-1 or 2-to-1 roping, or direct-traction machines with 2-to-1 or 3-to-1 roping.

It is in their car that passenger and freight elevators differ most. Where passenger-elevator cars can be objects of beauty, freight-elevator cars must be designed to carry the heavy loads they have to handle. These cars not only have to support their rated loads safely, but they must also resist the heavy side thrusts and twisting forces caused by loading and unloading, particularly with power trucks. To do this the cars and their slings are built of heavy steel construction, of a size to accommodate their loads. Also, guide rails must be fastened to the building structure with heavy steel brackets at closely spaced intervals to keep rail deflection within safe limits.

Where normal spacing of structural steel beams around the hoist-way is not close enough, vertical structure members may be installed

Fig. 1-34. Trucking sill S in the use position installed on a freight-elevator car platform where it bridges the opening between the car and building floor landing.

to provide additional support for the guide rails. Loading and un-loading forces and wear and tear on the car and guide rails can be greatly reduced by keeping the car level with the floor when heavy loads are being moved on or off, as explained on page 21. This is a big factor in favor of using variable-voltage control with automatic leveling to compensate for hoist-rope stretch or other causes of small movements of the car from floor level.

Trucking Sills. At best, moving a heavily loaded power truck on or off an elevator-car platform is not an overly smooth operation, even if the car is level with the floor. When the car is not level with the landing floor almost anything can happen to the truck and its load, and has. Besides damage to the truck and its load, hoistway doors may be rammed. To eliminate these difficulties, door-ramp sills are used extensively. These may be mounted on the car platform or on the floor landings. When mounted on the car, one sill serves all floors, but when landing-mounted a sill is required for each floor served by the elevator.

Figure 1-34 shows an Alexander trucking sill S in the use position, mounted on a freight-elevator car platform, where it bridges the opening between the car and building-floor landing. Counterbalanced by a weight W in a steel housing, the sill is manually operated by a recessed plunger-type crank handle mounted near the elevator control panel. Each sill has two electric contacts and a mechanical lock to prevent elevator operation unless the sill is locked in the raised position on the elevator platform. While the elevator is running the sill is locked in raised position to prevent accidents caused by trucks rolling against the car gate or the hoistway doors. With these sills it is not necessary to have the car platform exactly level with the floor landings for smooth operation of power trucks going on and off the car platform.

Sills mounted on the car, even if they are low in cost, are recommended only for freight elevators that are operated from inside the car. They are not best suited for automatically controlled freight elevators operated from landing push buttons. For this service a trucking sill should be mounted to the building-floor landings of each elevator door opening. They are recommended for all freight elevators equipped with motorized or manually operated hoistway doors and car gates. Elevator service is speeded up and operation made safer because the sills function automatically with operation of the doors and gate.

Garage Elevators. All the earlier multistory garage elevators were freight designs. In garages with only one elevator, cars could not be stored or delivered faster than the elevator could handle them, which frequently was unsatisfactory. Many garages are built with ramps, on which the cars are driven to or from the various floors. These too are unsatisfactory, because of the limit to the building height, about five stories, in which ramps can be used, high cost of operation, and other factors.

Within the last decade much study has been given to off-street car parking in downtown areas and systems of semiautomatic and automatic parking in multistory buildings have been developed. Among the semiautomatic types in use are Bowser and Pigeon Hole. For these systems the garage storage space, which may reach a height of

14 stories or more, is built on two sides of an elevator runway that extends the full height and length of the building (Fig. 1-35). On each side of the elevator runway the floors are divided into stalls one or two cars deep (Fig. 1-36). The runway is divided into one to five sections, each of which has an elevator that can travel vertically and horizontally.

FIG. 1-35. Elevator car *C* operates in a portable hoistway *H* suspended from a crane bridge to move the car horizontally and vertically in the elevator runway, which extends the length of the garage.

Each elevator car *C* runs in a structural steel hoistway *H*, which along with the elevator machine is carried overhead on a specially designed crane. This travels on overhead rails and is driven by a worm gear, so that the elevator machine, car, and its hoistway can travel horizontally while the car is moving vertically. As a result the car can travel diagonally in the runway. An antisway system of ropes attached to its bottom keeps the hoistway vertical at all times when it is in motion.

With this system cars are attendant-stored and -delivered. When a car is driven to the garage an attendant takes it over, gives the driver a receipt, and drives the car onto an elevator. From where he is seated in the car, the attendant presses a button for an empty stall indicated by a light on a panel on the side of the elevator car (Fig. 1-37). The elevator starts and goes directly to the stall at 400 fpm, if necessary traveling vertically and horizontally simultaneously.

When a customer comes for his car, he presents his receipt to the cashier, who pulls its corresponding stub from an electronic rack in his office. Removing the stub lights a lamp in the elevator which signals the attendant, who presses the button below it, the elevator going to the proper stall automatically. The attendant drives the

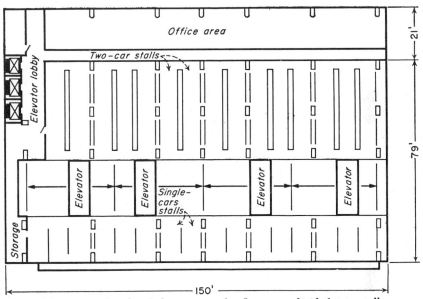

Fig. 1-36. On each side of the runway the floors are divided into stalls one or two automobiles deep.

car on the elevator, lowers it to the ground floor, and delivers it to the waiting customer.

These elevators are provided with all the safety devices normally found on a modern freight elevator. Gates of expanded metal welded to a steel frame, on the car and on the hoistway at the floors in front of each parking stall, must close before the elevator can start. These gates open automatically after the car has arrived and stopped at the stall to which it was signaled.

The Pigeon Hole–type semiautomatic garage differs from the Bowser in that it has lever control operated by the attendant from the car floor. It also has a car loading and unloading device to move the cars on and off the elevator. The elevator's movable hoistway is supported from the bottom and moves horizontally on steel rails. Both systems accomplish the same purpose, parking automobiles in stalls on each side of a movable elevator hoistway and retrieving them when they are wanted.

Automatic Parking. Two of these systems are Park-O-Mat and Alkro, each using elevators in fixed hoistways to park automobiles

Fig. 1-37. The attendant, from where he is seated in an automobile on the elevator car, presses a button for an empty stall indicated by a light on a panel on the side of the car.

from either end of an elevator car. With this system an attendant does not ride on the elevators, but remains at a push-button control station on the first floor. From here he can, by pressing buttons, operate two or more elevators for parking or delivering cars automatically. To make this possible the cars are handled on automatic positioners for parking and delivery. Another system of automatic parking, Speed-Park, was developed by Speed-Park Inc., in cooperation with Otis Elevator Company.

For further details on garage parking systems see *Elevator World*, August and September, 1954; February, 1955; July, 1956; and March, 1957.

Dumbwaiters. The elevator shown in Fig. 1-30 is representative of the hookup for low-rise electric-powered dumbwaiters, which may be considered small electric elevators. They are used in stores, restaurants, hotels, hospitals, and other buildings to handle food, packages, and other supplies. Many of them serve only two or three floors. These are usually driven by a drum-type machine, like that shown in Fig. 1-30, but of smaller capacity, about 1.5-hp rating, located in the hoistway below the car. Above three-floors' rise, dumbwaiters may be driven by traction machines.

Electric dumbwaiters are operated by push buttons at each landing. Two-landing designs usually have two buttons at each landing and are wired for call and send operation. With this control the car can be sent from either landing. Three-stop installations are usually arranged for multibutton operation with three push buttons at each landing. One of the three buttons will call the car from either of the other two landings and the other two buttons are for sending.

Cars may be from 2 to 3 ft square by from 2 ft 6 in. to 4 ft high and built of wood or light steel, divided for one or more trays. Since no one can ride on the car, it does not generally have an overspeed governor and car safeties, but can be equipped if required. Operation of the car is foolproof so that it cannot be operated while a hoistway door is open and a door cannot be opened unless the car is at the landing. Other safety devices are provided to prevent damage to the equipment or injury to attendants.

Home Elevators. Personal-service elevators are available for home use in many forms. The most complete of these are similar to elevators found in apartment buildings but on a reduced scale. Modern ones have a traction machine and counterweights and the cars have instantaneous-type safeties operated by broken-rope devices or overspeed governors. These machines are available for a maximum rise of 35 ft at a speed of 20 to 35 fpm and a rated lifting capacity of 350 to 700 lb.

Operation is automatic from one push button at each landing and one button in the car for each landing served. Also in the car are an alarm button, a stop switch, a light switch, and frequently a telephone. Door interlocks and contacts prevent hoistway doors' being opened if the car is not at the landing and prevent the car's being started if the hoistway doors and car gate are not closed. Operation is on alternating current of any standard voltage and frequency including 110-volt, single-phase.

At the other end of the scale are designs that permit locating the car in an open stair well (Fig. 1-38), a closet in a corner of a room or hallway, or within a hoistway. Car platform may be any size or shape to fit the available space, up to a size to accommodate a passenger and a wheel chair with an attendant. No overhead construction is required as the car travels along a single steel channel enclosed in the form of a post, shown at *C* in Fig. 1-38.

The power unit, a motor-driven back-locking worm gear, is located in the basement or elsewhere out of sight, and is operated on house

Fɪɢ. 1-38. Home elevator, located in open stair well, may also be installed in a closet or in the corner of a room.

current. Full electrical control is provided, including automatic stops, operating buttons in the car, and send and call buttons at the landings. Safety devices, such as a back-locking worm-gear power unit, safety catches, slack-rope and overrun switches, and electrical and mechanical interlocks on all gates and doors protect against every operating hazard.

When the car descends in an open room or stair well, a safety stop diaphragm under it gives complete protection. Any slight pressure on the diaphragm immediately stops the car before its weight can rest on anything underneath. Where the car runs through an opening in the second floor this opening is protected by a floor door, white plate (Fig. 1-39). This door is always closed except when the car comes up through the floor. It then lifts the door on its top as in Fig. 1-40. In the rear (Fig. 1-39) is the channel that guides the car, and in which the hoist rope and the overhead sheave can be seen.

Stair Elevators.[1] Frequently the answer to the private-home elevator for those unable to climb stairs is a stair or inclined elevator (Fig. 1-41). These elevators are also called stairlifts, stair-climbers, Inclinators, Wecolators, Escalifts, stair-travelors, stair-glides, and stair-chairs. On the design shown in Fig. 1-41, the chair or car is supported on a roller truck running within a special steel channel

Fig. 1-39. Car travels along a single steel channel in background.

Fig. 1-40. Car lifts floor closure as it comes up through the floor.

secured to the stairs, generally along the wall. When not in use the chair folds back to the wall so that the chair and the channel take very little space and do not interfere with normal use of the stairs. Installation is simple and does not require cutting or defacing the building.

Several different methods of powering the chair are used. On the one shown in Fig. 1-41, a roller truck connects to a light steel rope that runs inside the channel to the top of the stairs, then over

[1] Information on stair elevators was supplied by the Inclinator Company of America, Shepard-Warner Elevator Company, F. G. Arwood & Company, W. E. Cheney Company, Sedgwich Machine Works, and American Stair Glide Corporation.

Fig. 1-41.   In this stair elevator the chair is supported on a roller truck running within a steel channel.

a sheave to a power unit, in the basement (Fig. 1-42), or in an out-of-the-way place in the home.   This unit consists of a ⅓-hp single-

Fig. 1-42.   Stair-elevator power unit located in basement.

phase 110-volt motor driving a self-locking worm and gear and a drum on which the rope winds.   Operation is from constant-pressure push buttons on the chair and a call button at each landing.   Limit switches automatically stop the car at the end of travel if the passenger holds the button closed.   Other safety devices such as self-locking gears and slack-cable switch protect against any hazard that might develop.

In other designs of stair elevators the power unit is built in as part of the chair.   This also consists of a self-locking worm and gear, electric brake, and direct-connected motor. Some of them drive the car by a gear meshing into a rack gear, fastened to the bottom of the steel or aluminum channel that supports the chair. Others drive through a fixed steel roller chain fastened at both ends of the track. A sprocket driven by a small worm and gear and a ¼-hp motor travels up and down the chain. Two spring-loaded deflector sprockets on the carriage permit operating a slack-chain switch and a broken-chain safety catch.

This carriage has eight nylon roller bearings and travels on an extruded-aluminum channel track equipped with stair attachment brackets arranged so that the track may be adjusted to suit any stair incline. In addition to carrying the power unit, the carriage also

Fig. 1-43. In this stair elevator the power unit is built in as part of the chair, which is propelled by a traction drive.

provides a base for the chair, adjustable to suit the stair incline. On this unit safety devices include a broken-chain safety catch, slack-chain cutoff switch, travel limit switches, magnetic brake on the worm shaft, constant-pressure push-button operation, and thermal overload protection of the driving motor. Another stair elevator, shown in Fig. 1-43, has a traction drive that uses heavy rubber-tired rollers to propel the chair and cushion its ride. No racks, cables, or chains are used.

All stair elevators are built to operate in straight stairways and some can be installed to run in curved ones. If the stairways are continuous some stair elevators can operate through two or more stories. Stair elevators are generally rated 250 to 300 lb at 25 to 35 fpm on a maximum incline of 45°. Their cars are usually built for one chair and passenger, but some have facing chairs for two persons (Fig. 1-41).

Alternating Current. Where only alternating current is available in a building there are several ways of operating elevators. When a building has been supplied with direct current and changed over to alternating current, rectifying equipment can be installed to supply direct current. Then, nothing has to be changed but the primary power supply. This has been done in some large buildings by installing large mercury-arc rectifiers.

In other buildings all equipments have been changed to a-c power but the elevators. These remained on direct current supplied through rectifiers, which are frequently of the metallic-dish type, such as copper oxide, selenium, or germanium designs. Such installations are usually limited to medium-size and smaller buildings and with their control form a single assembly. These can be moved into the building and connected to the power system over a weekend, during the time a changeover is made from direct to alternating current. In such installations provisions may have to be made to take care of regeneration from the elevator motors to prevent overspeeding in the down direction; see page 151.

Another method of converting alternating current to direct current is by a specially designed motor-generator set used with a variable-voltage (Ward-Leonard) control system. Such a system is used by practically all elevator manufacturers who build high-speed automatic and nonautomatic elevators. In it the motor-generator set is part of the control and generally consists of a squirrel-cage motor driving a d-c generator and exciter. Usually one motor generator is used for each elevator motor. Speed and direction of elevator motor rotation are controlled by adjusting the current in the generator-field windings as described in Chap. 9. This system may also be used on worm-gear traction machines for car speeds below 400 fpm.

Squirrel-cage Motors. These are used extensively on slow- and medium-speed elevators in a number of motor and control combinations. However, none of them is applicable to high-speed elevator service, such as obtained with modern direct-traction machines. When this class of service is required, involving car speeds above 400 fpm, variable-voltage control with a d-c motor generally offers the best solution to the problem. Alternating-current motors have been used for car speeds up to 700 fpm but are now generally limited to 200 fpm or less.

Because of their comparatively low starting torque, general-purpose squirrel-cage induction motors are not suited to elevator service. Experience shows that on a worm-gear elevator a motor must develop 200 per cent as much torque to start and accelerate the machine with full load in the car as is required to hoist the load. Therefore, to be on the safe side, a-c squirrel-cage motors are designed to develop about 250 per cent rated full-load torque. These high-torque motors under light-load conditions will accelerate the car so quickly that the passengers may be given an unpleasant jolt. To overcome this difficulty it is general practice to connect at least one step of resistance in series with the stator winding to reduce the motor's torque about one-half at the instant of starting. After the motor starts the resistance is cut out of circuit and regardless of loading the elevator starts smoothly.

Wound-rotor Motors. These have been used on elevators in the past with controls that cut out the rotor starting resistance in a number of steps. Today high-torque squirrel-cage motors are used to eliminate the complications and high cost of wound-rotor motors and their controls. Also, such motors cannot be slowed down below rated speed when stopping the elevator, no more than a single-speed squirrel-cage design can. This has led to a wide use of two-speed types.

Multispeed Motors. Although a good operator can land a car at speeds of 150 to 200 fpm, it is desirable to get the speed down to at least 60 fpm before cutting the power out of the motor. In freight service this may be as low as 10 fpm. On d-c machines the speed of the car can be taken care of in the control equipment and almost any desired landing speed obtained. With a-c motors, when a heavy load is lowered, connecting resistance in the stator or rotor circuit will cause car speed to increase instead of decrease. During such load conditions the motor is acting as a generator and is supplying power into the system to produce a braking action and keep the elevator under control. If resistance is connected into the circuit, the motor's speed will have to increase to supply the necessary current and braking effect to keep the car under control. For this reason the car must be landed from full speed if a single-speed motor is used. This condition puts a limit of about 50 fpm on car speeds for single-speed motors. For speeds higher than this, two-speed motors are generally used, which are usually of the squirrel-cage type. Some of the elevator manufacturers use multispeed motors for practically all car speeds.

Motors of this type are built in three different forms. First, they are built with a single winding arranged for regrouping to give two different speeds. These motors are limited to speed changes not exceeding one to four by regrouping the winding through suitable control equipment for two different numbers of poles, for example, 24 to 6 poles. On a 60-cycle circuit the 24-pole grouping would give the motor a synchronous speed of 300 rpm and the 6-pole grouping a synchronous speed of 1,200 rpm.

The second method of obtaining a two-speed motor is to wind the stator with two separate windings in the same slots. The windings

FIG. 1-44. Geared-traction passenger-elevator machine driven by a 6-to-1 two-speed squirrel-cage motor.

are grouped for a different number of poles, such as 24 and 6, which would give the same speed ratio as for the single-winding machine. Motors of this type have been built for speed changes of as high as one to six. For example, if on the slow-speed winding the motor runs 150 rpm, on the high-speed winding it operates at 900 rpm. Motors of this type have been built for operation on elevators having car speeds of 700 fpm, such as the machine shown in Fig. 1-44.

A third type of two-speed motor consists of two separate motors in the same frame (Fig. 1-45). One of the stator cores has a winding grouped for, say, 16 poles, and the other, 4 poles. Such an arrangement gives a speed change of one to four. Motors of this type are limited in speed range by the same factors as the double-winding motors.   In the double motor the two rotors are keyed on the same shaft without a coupling.

Starting Two-speed Motors.   Normal practice does all starting on the high-speed winding and uses the slow-speed winding only for stopping.   At starting, a resistor is connected in series with the high-speed winding; it is then cut out as the motor accelerates.   This gives smooth acceleration up to full speed without changing from one winding to the other.   Slowdown is accomplished by changing to

Fig. 1-45.   Traction-elevator machine driven by a two-speed double squirrel-cage motor.

the slow-speed winding with a resistor in circuit while the high-speed winding is still connected.   The high-speed winding is then disconnected and the slow-speed winding's resistor short-circuited.   A smooth dynamic-braking action is obtained that slows down the elevator to a speed corresponding to the motor's normal slow speed. After the motor is disconnected, the mechanical brake is applied to stop the elevator.   This method greatly reduces the work that must be done by the brake and it can be much smaller than if it had to stop the elevator from full speed.   It also permits reducing the car speed to a speed at which a good landing stop can be made.

Squirrel-cage motors, either single- or two-speed, seldom overheat when applied in normal elevator service, where the stops do not exceed 150 per hr including jogging. When the starts per hour are excessive, as they may be with department-store elevators, and when car speeds are comparatively high, a-c elevator motors may overheat. This is because they must dissipate internally accelerating as well as full-speed losses.

Forced-ventilated Motors. For elevator power requirements that cannot be handled by standard motors, special forced-ventilated two-speed squirrel-cage designs may be used with a separate motor-driven fan, mounted on the main motor. They will handle, without

FIG. 1-46. Two-speed elevator motor includes a high-speed slip-ring induction unit combined with a slow-speed squirrel-cage type.

excessive temperature rises, 300 starts per hr when driving a load equivalent to normal elevator service. On some designs the ventilating motor always run when the main motor operates. However, a thermostat within the latter causes the ventilating motor to run as long as necessary after the elevator stops to keep the main motor's temperature within safe limits. On other forced-ventilated motors the ventilating fan runs continuously to provide full ventilation for the main motor during acceleration, running, slowdown, and standing time.

Another two-speed motor that has been used for severe elevator service comprises a high-speed slip-ring induction motor, mounted in tandem with a low-speed squirrel-cage motor (Fig. 1-46). The slip-ring motor serves for acceleration and full-speed running while the low-speed squirrel-cage provides slowdown and landing speeds. Since acceleration losses of the slip-ring motor are largely dissipated in an external resistor and its full-speed losses are comparatively low, it can handle a large number of starts per hour without overheating and was built for 4-to-1 and 6-to-1 speed ratios.

Single-phase motors have been used on elevators, where polyphase current was not available. These motors have generally been of the repulsion type.

A-C Motor Controllers. Many of the controllers on a-c elevators are operated by magnets in much the same way as on d-c machines, except that the magnet cores are laminated to reduce eddy currents and prevent excessive heating of the cores. At best, a-c controllers tend to be noisy, and cost more than corresponding d-c controllers. For these reasons, elevators driven by a-c motors frequently have d-c controllers and brakes. With this system, the elevator motor alone operates on alternating current; the electric brake and all contactor magnets and relays are actuated by direct current supplied by a metallic-disk rectifier, usually of the selenium, copper oxide, or germanium type; see Chap. 8. All disagreeable control and brake noises are eliminated with elevator performance smooth and quiet. Rectifiers have made it possible to obtain with alternating current many of the advantages of d-c operation of slow- and medium-speed elevators.

On some a-c elevator-motor controllers the magnets are immersed in oil pots to help keep them cool and quiet and ensure lubrication. Rotating types of magnets are also used. These are simple polyphase torque motors, controlled from the car switch or push buttons. Figure 1-47 shows a two-speed squirrel-cage motor controller of this type. The torque-motor-type magnets are on the left and have a constant pull over their entire range of action.

These magnets also take a constant current. Therefore, they do not have the large inrush common to other types. Because their core does not seal or close, the magnets are known as a nonsealing type. Polyphase nonsealing brake magnets are also used as on the machine shown in Fig. 1-44. This magnet, although of the polyphase revolving-field type, reciprocates in operation instead of rotating. When it is energized it lifts its core to release the brake, like other electromagnets. Operation of several type of brakes is discussed in Chap. 3.

Power Consumption. Power consumption of an elevator motor is affected by the type of machine, type of control, the number of stops per car mile, and other factors. Tests on five elevators for one year, averaging 180 stops per car mile and using rheostatic control, showed an average power consumption of 8.1 kwhr per car mile. With variable-voltage control and about 100 stops per car mile, the power consumption was about 4.5 kwhr per car mile.

Fig. 1-47. Two-speed squirrel-cage motor controller with torque-motor-magnet-operated contactors.

Using variable-voltage control in local service, making about 370 stops per car mile, the power consumption was about 7.5 kwhr per car mile. In the first instance, the machines had a rating of 2,500 lb at 600 fpm; in the second, 2,000 lb at 500 ft; and in the third, 3,500 lb at 350 fpm.

CHAPTER 2

Methods of Roping and Their Effects on Loading of Ropes and Bearings

Drum-type Machines. For a given weight of car and load the method of roping up the machine has a considerable effect on the loading of the ropes and bearings. Practically all earlier types of electric elevator machines were winding-drum types, but in present-day practice this type is being rapidly superseded by traction types. On drum machines there are generally three sets of ropes. One, the car ropes, has one end attached to the car and the other to the winding drum; a second set, the drum-counterweight ropes, runs from one counterweight to the drum; the third set connects the car to its counterweight. All these are shown in Fig. 2-1 for a basement installation.

The overhead drum elevator installation (Fig. 2-2) simplifies the roping considerably and requires only about one-half the length of rope from the car and from the counterweight to the drum. The car counterweight ropes are about the same length in either installation. The general arrangement of the ropes is shown in Figs. 2-1 and 2-2, where the car and drum-counterweight ropes lead off opposite ends and sides of the drum, with the car counterweight ropes passing over the vibrating sheave and to the counterweight. When the drum spans less than half the width of the hoistway, a deflecting sheave is required to lead the counterweight ropes vertically down the hoistway.

These machines are limited to relatively short rises of about 150 ft or less. For higher rises the drum becomes of an unwieldy length to provide space on which to wind the ropes. With a drum of given diameter a different length is required for each height of car travel. For example, on a 100-ft rise a drum would have to have space to wind 50 ft more rope than on a 50-ft rise.

Another objection to drum machines is that if the limit and control devices fail to cut the power out of the motor at terminal landings, the car or counterweight may be pulled into the overhead work. In such an event the ropes may be pulled out of their sockets and the machine wrecked.

With the overhead machine (Fig. 2-2) the ropes have to bend in only one direction when winding on and off the drum, a condition favorable to long rope life under normal conditions. When such machines are in the basement (Fig. 2-1), reverse bends are put in the ropes as they bend around the different sheaves, which tends to shorten rope life, because of individual wires breaking from fatigue.

In an overhead installation, because the ropes travel along the drum as they wind on or unwind off, there is only one point in the car and counterweight travel where they exert a vertical pull. At all other places in the hoistway a side pull is applied to the car and counterweights. The effect as applied to the car is indicated in Fig. 2-3, which shows that if the ropes pull vertically on the car at the top landing, at every other position there is a side pull. This is also true for the counterweights. To avoid this side thrust in high-rise installations, a single spiral groove is sometimes used on the drum, and one-half of the drum is grooved right-handed and the other left-handed. One of the two car ropes is attached to each end of the drum and the two counterweight ropes at the center, as in Figs. 2-4 and 2-5. From these figures it can be seen that although the ropes lead off from the vertical, one opposes the side thrust of the other and gives a vertical pull on car and counterweights at all positions of travel.

Drum Traction Machines. What might be considered as a transition from the drum elevator machine to the traction type is known as the drum-type traction drive. This machine uses a single spiral U-grooved drum. The ropes run from the car and are given $1\frac{1}{2}$ or $2\frac{1}{2}$ wraps around the drum; they then go to the counterweight (Fig. 2-6). The ropes do not fasten to the drum but depend on their friction in the grooves to hoist the car or counterweight. In their operation on the drum the ropes act as nuts having $1\frac{1}{2}$ or $2\frac{1}{2}$ threads on a bolt. As the drum turns the ropes travel from one end of it to the other as a nut travels along a bolt. Since the ropes bend in one direction only they have a long life under normal conditions.

This type of machine is subject to the same objections as the winding-drum type; namely, a side pull is applied to the car in all but one position in the hoistway, different lengths of drum are re-

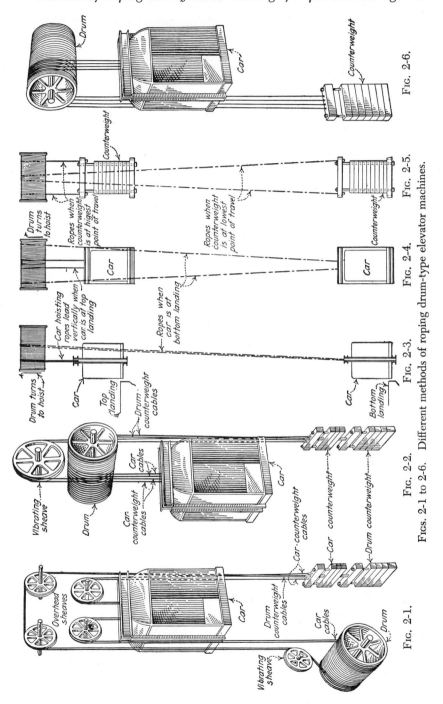

Fig. 2-6.

Fig. 2-5.

Fig. 2-4.

Fig. 2-3.

Fig. 2-2.

Fig. 2-1.

Figs. 2-1 to 2-6. Different methods of roping drum-type elevator machines.

quired for different heights of maximum car travel, and in high rises
the drum becomes long and unwieldy. A limit switch is provided
to stop the motor as the ropes approach the end of the drum. Although the machine cannot pull the car or counterweight into the
overhead beams, if for any reason the safety limits fail there could
be a bad piling up of the ropes at the end of the drum, which
might be serious.

Double-wrap Traction Machines. Traction-elevator machines, as
they are generally known, may be divided into two general classes,
double- or full-wrap and single- or half-wrap. In order to obtain
sufficient traction between the ropes and traction or driving sheave,
with U-shaped grooves, a secondary or idler sheave is used, as is
shown in Fig. 2-7. The sheaves have parallel U-shaped grooves instead of the spiral grooves used on the winding drum of drum machines. In Fig. 2-7 the ropes run from the car over the traction
sheave down around the idler sheave, up over the traction sheave
again, and to the counterweight. This gives the effect of a full wrap
on the traction sheave, from which this type of elevator gets the
name "double-" or "full-wrap."

Three to eight ropes are generally used, and the traction sheave
always requires twice as many grooves as there are ropes. To prevent excessive strain being put in the ropes, the sheaves should not
be less than 27 in. in diameter, or forty times the rope diameter.

With the traction type, a standard machine can be used irrespective of the length of car travel. If the car or counterweight lands
on its bumpers at the bottom of the pit, traction is lost between the
ropes and sheaves, so that neither car nor counterweight can be
pulled into the overhead work. Therefore, if the limit devices fail,
the car or counterweight lands and the motor runs and turns the
traction sheave in the ropes until the power is cut off.

Where traction-sheave diameter equals the distance between the
centerlines of the car and counterweight rails, the secondary sheave
may be placed directly under it, as in Fig. 2-7. Where this distance
is greater than the diameter of the traction sheave, the idler sheave
also serves as a deflector to lead the ropes vertically down to the
counterweight (Fig. 2-8).

Generally traction-elevator machines are located overhead. When
it becomes necessary to install the machine in the basement, roping
is similar to that shown in Fig. 2-9. This requires a somewhat similar arrangement of overhead sheaves as with the basement-type drum
machine and also causes reverse bends in the ropes. About twice
the amount of rope is required for this installation as for an over-

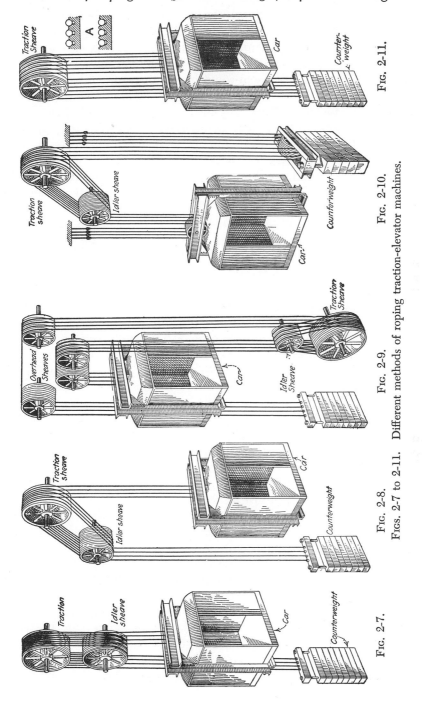

Fig. 2-11.

Fig. 2-10.

Fig. 2-9.

Fig. 2-8.

Fig. 2-7.

Figs. 2-7 to 2-11. Different methods of roping traction-elevator machines.

head machine, and additional sheaves make the cost considerably more than for overhead installation. Furthermore, rope life will be shorter and power consumption higher; consequently operating expenses will be greater and loading on the building may be increased. For these reasons basement-type installations should be avoided whenever possible.

Traction Machines, 2-to-1 Roping. In traction installations so far considered, speed of car and counterweight equals the peripheral speed of the traction sheave. This is what is called a 1-to-1 roping, in that the car and rope speeds are the same. Where it is desirable to use a direct-traction-elevator machine, for a car speed lower than can be practically obtained with a 1-to-1 roping, a 2-to-1 roping may be used, as shown in Fig. 2-10. This arrangement is also used on some classes of geared traction machines, where very slow car speeds are required, such as on heavy freight and automobile elevators.

With 2-to-1 roping car speed is only one-half that of rope speed. Instead of the rope ends being attached to the car and counterweight, as for 1-to-1 roping, they are dead-ended to the overhead beams. From one of the dead ends the ropes pass around an idler sheave in the car crosshead, around the traction and secondary sheaves, then around an idler sheave in the counterweight crosshead and to the second dead end (Fig. 2-10).

Since one-half the weight of the car and the counterweight is carried on the dead ends (Fig. 2-10), loading on the traction and secondary sheaves is only about one-half that for the 1-to-1 machine (Fig. 2-8). Consequently, a lighter construction can be used. From Fig. 2-10 it will be seen that the ropes make three reverse bends. This may cause shorter rope life for the 2-to-1 than for the 1-to-1 overhead installation.

Half-wrap Traction Machines. A type of traction machine coming into wide use is that known as the half-wrap, the roping scheme of which is shown in Fig. 2-11. In this type of installation the ropes pass from the car over the traction sheave to the counterweight, so that they make a half wrap only on the sheave. The traction sheave has some form of a wedge-shaped groove which grips the ropes by virtue of a wedging action between the sides of the grooves and ropes. Two types of the grooves are shown at A in Fig. 2-11.

This type of installation gives one-half the loading on the traction sheave that is obtained with the full-wrap machine for the same weight of car load and counterweight. Consequently, the former can be constructed considerably lighter than the latter. With one-half-wrap installation there are a minimum number of bends in the ropes, which helps toward long rope life. This, however, is offset to a considerable degree by the pinching action of the grooves on the ropes; but rope life may be a little longer on single-wrap than on double-wrap installations. For the same capacity and speed the same machine may be used for any height of car travel. If the limit switches fail to function, the car or counterweight will land on its buffer, when the traction between the ropes and sheave will be relieved and the ropes will not be strained or pulled out of their sockets, as is likely to happen on a drum-type machine.

As the grooves wear there is a tendency for the pinching action on the ropes to be reduced. This is not a serious factor in the machine's operation since sheaves have given 8 to 10 years or longer continuous service without experiencing any difficulty from wearing of the grooves. Elevator codes now require that sheave grooves be of a type on which traction will not be materially changed with wear.

On either a double-wrap or a single-wrap installation condition of the sheave grooves must be carefully watched. When one groove wears more than the other, the rope in the groove with the shortest periphery will have to slip to have the same speed as the other ropes. This slippage wears not only the rope, but its groove, and tends to make conditions worse. When grooves are worn so that their peripheries are not the same length, true them in a lathe. This work is sometimes done by mounting a lathe tool and rest on the machine and turning the sheave with the motor. If the sheave is too badly worn, it will have to be replaced.

When the traction-sheave diameter is less than one-half the distance between the centerlines of the car and counterweight rails, a deflecting sheave can be used to guide the ropes vertically down the hoistway to the counterweight, as with the double-wrap installations shown in Fig. 2-8. The limit for deflecting the ropes off the sheave is about 45° or less from the vertical. When sufficient span cannot be obtained in this way, such as with wide freight cars, it may be necessary to use a double-wrap installation.

There are other schemes that have been used to rope up elevator installations, such as a 3-to-1 arrangement, but those given in the foregoing have come into most common use and will probably include over 99 per cent of all electric elevators in use.

Load on Ropes without Counterweight on Drum Machines. The simplest form of a winding drum elevator machine is that shown in Fig. 2-12, which consists of a car of some kind connected by ropes to a winding drum. Objection to this arrangement is that the motor must be large enough to lift the total load, including the car. The ropes are subjected to the full load, and the brakes must be strong enough to stop the total load, including the car, in the down motion.

For example, if the car weighs 2,500 lb and the rated load is 2,500 lb, then the motor must be capable of operating the machine when lifting 5,000 lb. Loading on the ropes and overhead sheave would be as indicated. Ropes are subjected to the weight of the car load and the overhead-sheave bearings are loaded to double this value plus the weight of the sheave. If the drum is driven by a worm and gear, the rear thrust bearing will be subjected to a load nearly equal to the weight of the car and its load. When the car is stopped in the down motion, loading on the thrust bearing will greatly exceed the weight of the car and its load.

The great difference between stopping the machine without a load in the car and when fully loaded, especially in the down motion, makes this arrangement very unsatisfactory. Furthermore, in the up motion the motor must always lift at least the car, and in the down motion there will always be at least the weight of the car driving the motor as a generator and pumping back into the electric system, which results in an unnecessarily large power consumption.

Load on Ropes When Drum Counterweight Is Used. The roping arrangement shown in Fig. 2-13 overcomes some of the objections to the scheme shown in Fig. 2-12. Here a drum counterweight is used and its ropes attach to one end of the drum and the car ropes to the other. In the up motion the counterweight ropes unwind and the car ropes wind in the same grooves, whereas in the down motion the reverse action takes place. It is general practice to make the counterweight equal to the weight of the car plus 40 per cent of its rated load. Applying this rule to the car and load in Fig. 2-13 makes the weight of the counterweight equal to

$$2,500 + (2,500 \times 0.40) = 3,500 \text{ lb}$$

as indicated. Here the motor has to be large enough to lift a load equal to $5,000 - 3,500 = 1,500$ lb. This weight, 1,500 lb, is also the loading on the rear thrust bearing with full load in the car. Instead of the brake having to be large enough to stop and hold a load of 5,000 lb when the car is in the down motion fully loaded, it need only be of sufficient capacity to take care of an unbalanced load of 1,500 lb. With 1,000-lb load in the car it is about balanced by the counterweight, so that very little power is required to move it in either direction. In the down motion with full load in the car, the motor will only have to resist a load of 1,500 lb to prevent the ma-

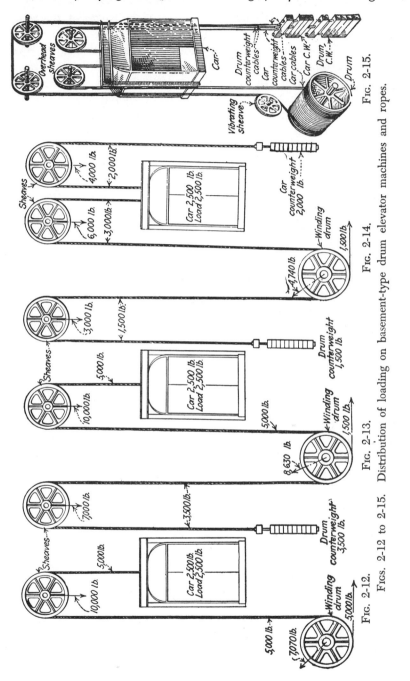

Fig. 2-15.

Fig. 2-14.

Fig. 2-13.

Fig. 2-12.

Figs. 2-12 to 2-15. Distribution of loading on basement-type drum elevator machines and ropes.

chine from racing. All these conditions help to make the machine more efficient in power consumption.

Adding the counterweight has not, however, reduced the loading on the car ropes, which is the same in both cases. On the other hand, the counterweight has increased the loading on the overhead beams 7,000 lb plus the weight of the additional sheave, making it 10,000 + 7,000 = 17,000 lb plus the weight of the sheaves. Loading on the drum-shaft bearings has been increased also from about 7,070 to 8,630 lb. This, however, does not take into account the downward loading of the drum or the upward thrust of the gear out of the worm. Instead of the drum-shaft bearing loading being vertical it is at an angle, due to the vertical lift of the car and counterweight ropes and the horizontal force transmitted from the worm and gear. Loading on the thrust bearing is based on the assumption that the worm gear is the same diameter as the drum. Where the gear is smaller than the drum, loading will be increased on the thrust bearing and the drum-shaft bearings.

Rope Loading When Car and Drum Counterweights Are Used. By using a car counterweight (Fig. 2-14), part of the car weight can be removed from the car ropes and drum and the weight of the drum counterweight can be reduced, which will also remove part of the load from the drum-shaft bearings. It is general practice to make the weight of the car counterweight from 300 to 500 lb less than the weight of the car. Making the car counterweight equal 500 lb less than the weight of the car gives a weight of 2,000 lb and reduces the drum counterweight 2,000 lb. Then, a 4,000-lb load would be taken off the drum shaft, besides removing a 2,000-lb load off the car-hoisting ropes. The loading on the thrust bearings will remain the same as in Fig. 2-13, as will the size of the brake, but the load on the overhead work has been reduced from 17,000 to 13,000 lb, a reduction of 4,000 lb. This reduction in weight on the overhead beams would be affected somewhat by the sheaves, but it is evident that the counterweight scheme (Fig. 2-14) has a distinct advantage over the arrangements shown in Figs. 2-12 and 2-13. It is for these reasons that the counterweights on drum-type elevators are generally arranged as shown in Fig. 2-14, although in practice the two sets of counterweights are generally arranged to run in the same guide rails, with the car counterweight above the drum counterweight, as in Fig. 2-15.

Overhead Drum Machines. At first thought it might appear that placing the elevator machine overhead would cause a much heavier loading on the overhead beams than locating the machine in the basement. Consideration of Fig. 2-16 will show that this is not necessarily so. For example, with the same load, weight of car, and counterweight, the load on the overhead beams from the car and counterweight is only 8,500 lb in Fig. 2-16, as against 13,000 lb plus the weight of the sheaves in Fig. 2-14, or a difference of 4,500 lb plus the weight of the sheaves. Therefore, if the machine shown in Fig. 2-16 weighs 4,500 lb, the loading on the overhead beam is about

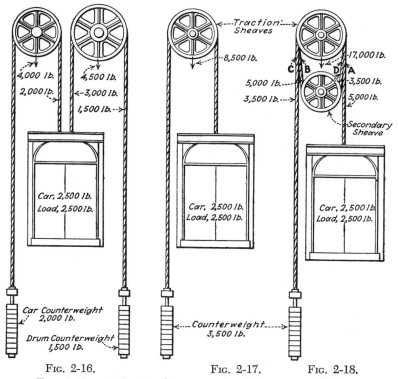

Fig. 2-16.   Fig. 2-17.   Fig. 2-18.
Figs. 2-16 to 2-18.   Loading on overhead elevator installations.

the same in both cases. In some installations, loading on the overhead beams is less with the machine placed overhead than it would be with the machine in the basement.

Traction-elevator machines are usually placed overhead (Fig. 2-17). In the simplest type, the ropes pass from the car over the traction sheave and to the counterweight. On the traction sheave, V-shaped grooves are used to provide sufficient traction between the ropes and the sheave to prevent the ropes from slipping. With full load in the car the maximum strain on the ropes is equal to the load and car weight or, in Fig. 2-17, 2,500 + 2,500 = 5,000 lb. The loading on the sheave bearings is the combined weight of the car load and counterweight, or 8,500 lb. This is 4,000 lb more than on the drum-shaft bearings of a drum machine installed overhead.

The machine shown in Fig. 2-17 is known as a V-groove or single-wrap type. With the double-wrap machine (Fig. 2-18), the ropes pass from the car over the traction sheave, down around the secondary sheave, back over the traction sheave, and to the counterweight. Neglecting the weight of the additional sheave and extra weight in the machine, load on the overhead beams shown in Fig. 2-18 is the same as in Fig. 2-17. However, loading on the traction-sheave bearings has been more than doubled. In Fig. 2-18, rope section *A* supports the car and its load, 5,000 lb. Rope section *B* may be considered as having a tension of 5,000 lb. Rope *C* sustains the load of the counterweight, 3,500 lb, and, to put this load into equilibrium, rope *D* may be considered as supporting a load of 3,500 lb. This gives double the loading on the traction sheave that exists in Fig. 2-17, although the friction between the ropes and the sheave will affect the loading in rope sections *B* and *D* so that it may not be exactly as shown in the figure, where an equivalent loading exists. Since the loading on the traction sheave in the double-wrap machine is double that for the single-wrap, for the same car load, the former must be constructed considerably heavier than the latter. Besides, a secondary sheave is required, all of which tends to make the double-wrap machine considerably more expensive than the single-wrap. With the double-wrap machine, friction due to the load on the main driving-sheave shaft and the idler-sheave shaft is three times as great as with the single-wrap machine.

In the foregoing, to simplify the problem, all the factors that affect the loadings have not been considered, such as those forces that develop in starting and stopping, but those that have been taken into account are sufficient to indicate the advantages of one method of counterweighting or roping over another. Although all the various schemes of roping up elevator machines have not been considered, the reader may work out the loading on any other arrangement by the simple process given. The weight of the car has been taken as 2,500 lb, but this will vary over wide ranges, depending upon design. It is not uncommon for passenger cars to weigh 5,000 lb and, in some special cases, probably 10,000 lb. These weights are for cars used on machines having a rating of 2,500 to 4,000 lb.

Why Compensating Ropes Are Used.[1] Without any form of compensation the load on an elevator machine or motor will not be con-

---

[1] The material on compensating cables was written by A. A. Gazda, Manager, Engineering Department, Kaestner & Hecht Company.

stant during a run from the hoistway's bottom to its top. Starting at the first floor (Fig. 2-19), the motor must not only lift the difference in weight between the car and the counterweight, but must also raise the weight of the ropes included between the limits of travel. As the car approaches the top floor (Fig. 2-20), the effective

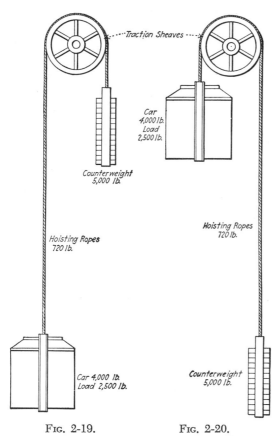

FIG. 2-19.   FIG. 2-20.

FIGS. 2-19 and 2-20.   Traction-elevator machines without compensating ropes.

weight of the ropes is transferred to the counterweight side, with the result that the load on the motor is reduced by twice the weight of the ropes. In other words, the weight of the ropes, which is added to the weight of the car and load in Fig. 2-19, is added to the counterweights in Fig. 2-20. With a rise of only a few floors this change in load is not important enough to warrant any special consideration. On the other hand, if a rise of 100 ft or over is involved, this factor becomes quite appreciable, and means to compensate for it should be supplied.

Methods of Rope-weight Compensation. The simplest method of elevator-rope compensation consists of attaching one end of the compensating ropes to the underside of the car and the other end at a

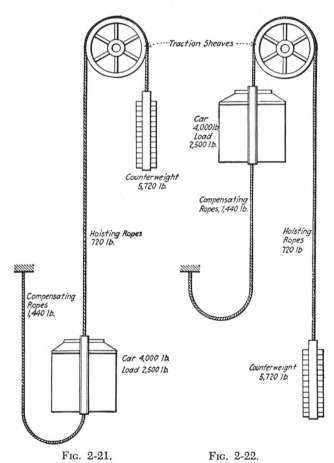

Fig. 2-21.   Fig. 2-22.

Figs. 2-21 and 2-22. Compensating ropes connected from car to center of hoistway.

point halfway up the hoistway. This is seen in Figs. 2-21 and 2-22, which show the car at the bottom and top of the hoistway, respectively. A variation of this method is given in Figs. 2-23 and 2-24, where the compensating ropes are divided into two parts, one being attached to the car and the other to the counterweight, while the free ends of each are fixed at the middle of the hoistway. Another method consists of using one set of ropes, with one end attached to the car and the other to the counterweight, as in Figs. 2-25 and 2-26. Before proceeding to the relative advantages of the three methods, a typical problem with its solution for each method will be considered.

Effects of Compensating Ropes. Assume a load of 2,500 lb in a car weighing 4,000 lb. The counterweight should balance the weight of the car plus 40 per cent of the load, giving a total of 5,000 lb for the counterweight. A conservative design for the hoisting ropes would specify six ⅝-in. steel ropes. These weigh approximately 0.6 lb per running foot. Hence, if the total travel is assumed to be 200 ft, the weight of unbalanced rope would be 720 lb. Turning to Fig. 2-19, it is seen that this will add a load of 720 lb on the motor at the first landing and subtract a similar amount at the top (Fig. 2-20). In Fig. 2-19, on the car side of the sheave the load is $2,500 + 4,000 + 720 = 7,220$ lb, and the load on the counterweight side 5,000 lb, or a difference of $7,220 - 5,000 = 2,220$ lb. When the car is at the top of the hoistway, as in Fig. 2-20, the weight on the car side of the sheave is $2,500 + 4,000 = 6,500$ lb, where on the counterweight side the load is $5,000 + 720 = 5,720$ lb, a difference of $6,500 - 5,720 = 780$ lb. Thus a total variation in load of $2,220 - 780 = 1,440$ lb will be imposed on the motor. Referring to the car and load specifications, the net load on the motor, neglecting the ropes, is 1,500 lb when hoisting the maximum load.

Advantages and Disadvantages of Different Methods of Compensation. The first method of compensation, to determine the weight of the compensating ropes and the additional counterweight required to keep the load on the motor constant at 1,500 lb when the car is fully loaded, as in Figs. 2-21 and 2-22, will be taken up. In Fig. 2-21, since one end of the compensating ropes attaches to a point midway up the hoistway, their weight will have no effect on the motor when the car is at the bottom of the hoistway. Therefore, the counterweight will be made up of three items, namely, weight of the car, 40 per cent of rated load, and weight of the ropes, or $4,000 + 1,000 + 720 = 5,720$ lb. The weight of the car, load, and ropes with the car at the bottom landing is 7,220 lb. Therefore the motor load is $7,220 - 5,720 = 1,500$ lb.

When the car is at the top floor, the load on the car side of the sheave is the weight of the car and load and equals 6,500 lb. On the counterweight side the load is the weight of the counterweight plus the weight of the ropes, or $5,720 + 720 = 6,440$ lb. Consequently, the load on the motor is only $6,500 - 6,440 = 60$ lb. If the load were taken off the car at the top floor, the weight of the car (4,000 lb) would be counterweighted by a weight of 6,440 lb. Therefore, for the motor to lower the car it would have to lift a load at starting of $6,440 - 4,000 = 2,440$ lb instead of 1,500 lb if the weight of the hoisting ropes were properly compensated. To main-

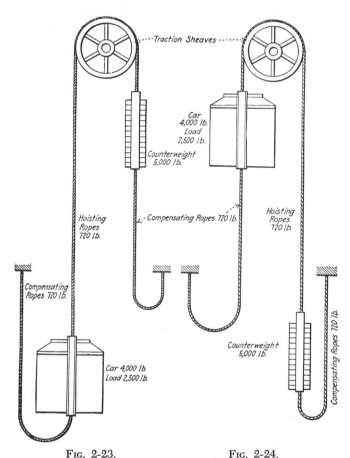

FIG. 2-23.   FIG. 2-24.

FIGS. 2-23 and 2-24.   Compensating ropes connected from car and counterweight to halfway up the hoistway.

tain the load on the motor at 1,500 lb would require attaching compensating ropes equal to $1,500 - 60 = 1,440$ lb to the car. In other words, the total weight of effective compensating ropes must be equal to twice the weight of the hoisting ropes included between the limits of travel.

Take the case of two separate sets of compensating ropes attached to the car and counterweight, as shown in Figs. 2-23 and 2-24. With the car at the bottom of the hoistway and the counterweight made equal to the weight of the car and 40 per cent of the rated load, the load on the motor when compensating ropes are not used is $7,220 - 5,000 = 2,220$ lb. To reduce this load to 1,500 lb would require compensating ropes attached to the counterweight, weighing $2,220 - 1,500 = 720$ lb. When the car is at the top landing, the weight of the hoisting ropes has been transferred to the counterweight, so that the load on the car side of the sheave is 6,500 lb and on the counterweight side $5,000 + 720 = 5,720$ lb. Therefore, without compensating ropes on the car the load on the motor would be $6,500 - 5,720 = 780$ lb. If the load is to be maintained at 1,500 lb, the car compensating rope will have to weigh $1,500 - 780 = 720$ lb. This shows that the same weight of compensating rope has to be used in either Figs. 2-21 and 2-22 or 2-23 and 2-24, but that in the latter case the counterweight can be reduced 720 lb.

Figures 2-25 and 2-26 show a direct car-to-counterweight compensation. With this scheme the weight of the compensating ropes is transferred to the counterweight when the car is at the bottom landing and to the car when the counterweight is at the bottom of the hoistway. In Fig. 2-25 the car, load, and ropes weigh 7,220 lb and the counterweight 5,000 lb, or a difference of $7,220 - 5,000 = 2,220$ lb. To reduce this difference to 1,500 lb requires compensating ropes attached to the counterweight of $2,200 - 1,500 = 720$ lb, which is the weight of the hoisting ropes between the limits of travel. When the car is at the top of the hoistway, 720 lb of the hoisting ropes have been transferred to the counterweights and an equal weight of compensating ropes to the car, so that the load on the motor is still 1,500 lb. This shows that not only is the counterweight at a minimum value of 5,000 lb, but the weight of compensating ropes is one-half that required with either of the other two methods.

Summary. A brief inspection of these three methods of hoisting-rope compensation, as shown by the foregoing discussion, will bring out the following points:

1. Method 1 requires the same weight of ropes as method 2, but additional counterweight, equal in weight to the hoisting ropes, must be provided. However, this method has the advantage that it can be used in practically any installation, while the others may be limited because of local conditions.

2. In method 2 space must be left for the second set of compensating ropes, which again depends upon local conditions. The ad-

vantage of this method is that the counterweight compensating ropes assist the counterweight, which may be held down to the minimum of 5,000 lb. This saving of 720 lb weight may offset the increased cost due to dividing the ropes.

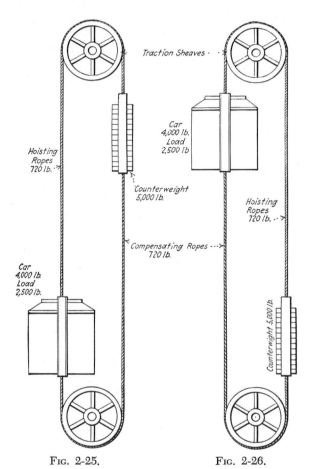

Fig. 2-25.   Fig. 2-26.

Figs. 2-25 and 2-26. Compensating ropes connected from car to counterweight.

3. Another factor that affects the design of the sheaves, particularly their bearings, is the total weight that will be suspended. In case 1 this will be 14,380 lb, while in case 2 it will be only 12,940, which shows a reduction of 10 per cent.

4. If the compensating ropes are attached to the car and counterweight as in case 3, they will transfer their weight in the same way as the hoisting ropes do. Hence, as shown by the calculations, only

one-half the weight of compensating ropes of cases 1 and 2 will be required. At the same time the counterweight is kept at 5,000 lb while the weight on the sheaves is 12,940 lb, as in case 2.

Thus it is seen that the third method is the best from the viewpoint of weight that must be handled. As far as accuracy of compensation throughout the travel is concerned, there is little choice between the three systems provided that they are all well designed. Sometimes it is impossible to apply the third method, as for instance if the counterweights are hung in a separate well and interference with other machines prevents the carrying across of compensating ropes. Another factor that is of importance in high-speed equipments is the elimination of noise. It is practically impossible to get noiseless operation using chains at speeds above 500 fpm. For this reason wire ropes are used in the application of the third method, even though the initial cost is increased.

CHAPTER 3

Overspeed Governors and Car and Counterweight Safeties

Safe Operation. Even though passenger and freight elevators operate in a vertical direction and the car and counterweight are supported from the ends of wire ropes, they are the safest means of transportation. This is because safety is designed into every part of them. With modern elevator control in good working condition the car cannot be started unless it is safe to do so. If the car should go out of control in down direction and overspeed, or the ropes break and let the car fall, an overspeed governor sets safeties on the car and locks it to the rails. In Chap. 1 safety devices were described in a general way. Here they will be given detailed study.

Compression-type Safeties. On medium- and high-speed cars, gradual or compression-type safeties are generally used, one design of which is shown in Fig. 3-1, installed in the safety plank under the car. *A* and *A* are steel channels bolted to safety heads *K*. Safety-cable drum *B* is keyed at *C* to drawbars *F* and has a cylindrical extension with holes *D* for releasing the safeties after they have been set by the governor, in case of overspeed. Free ends of the drawbars are threaded right and left and screw into heavy wedges *G*. Safety levers *I* pivot at *J* on heavy pins and have rollers *H* in their inner ends. Pinned to the outer ends of the levers are shoes *L* with cross-grooved surfaces for gripping the guide rails when the safeties are applied. Bumper plates *P* have been cut away to expose the safety levers. Holes *R* in the channels are for bolting on the bottom guide shoes.

One end of tail rope *S* connects to the drum on which the rope winds and the free end connects to the governor rope. Figure 3-4 shows diagrammatically a compression-type safety hookup with its governor at *A* and rope-gripping device *C*. Governor rope *B* is made endless by the double-end socket *E* held in releasing carrier *D* bolted to the top of the car. A weighted sheave *J* in the pit slides in guides *L* to hold a fixed tension in the governor rope. Tail rope *H* from the safety drum under the car comes up around sheave *I* to socket *G* and a stop ball *F* clamped to the governor rope.

The safety governor releases gripping jaws *C* to stop downward motion of the rope when the car reaches an overspeed of from 20 to 40 per cent of normal, depending on the rated speed of the car. The higher the rated car speed the lower the percentage of overspeed permitted before the governor trips. When jaws *C* grip the governor rope, downward motion of the car pulls the rope free of the releasing carrier *D*. Further movement of the car pulls the tail rope off drum *B* (Fig. 3-1). This action turns the drum and the drawbars to screw the latter into wedges *G* to pull them between rollers *H*. Safety levers *I* are spread apart at their inner ends and, rocking on pivots *J* forces shoes *L* against the guide rails. Pressure on the

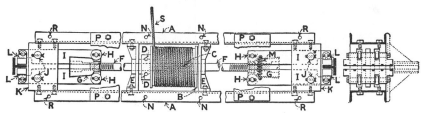

Fig. 3-1. Parts and assembly of a compression-type safety.

guide rails increases until friction between shoes and rails brings the car to rest. There are several designs for these safeties, but that in Fig. 3-1 shows their general construction and operation.

How to Release Safeties. To release the safeties on the car of a drum-type machine, the car-hoisting ropes are replaced on the machine's drum and the slack taken out slowly until they support the car. The governor gripping jaws are released and latched. Then a small opening in the car's floor is uncovered and a short round bar in holes *D* (Fig. 3-1) is used to turn the safety drum. This winds the tail rope back on its drum and forces the wedges from between the safety jaw levers. When the wedges are screwed back to their normal positions, springs *M* pull the safety shoes free from the guide rails.

After the safeties have been released, the connection on the governor rope to the releasing carrier should be restored and the governor inspected to determine if it is in good working condition and whether or not its rope was injured by the gripping jaws. The car should then be run down to the pit, the safeties checked from beneath the car, and the tail rope checked to see that it has been properly wound on its drum. Also, the rails should be inspected where the safety jaws gripped them to check whether or not their surfaces have been roughened and if they have to be smoothed with a file.

The safety shown in Fig. 3-2 operates like the one in Fig. 3-1, but is released by a tool with a gear on it that engages gear G. This tool is also used through a hole in the floor of the car.

Rope-applied Safeties. The safety shown in Fig. 3-3 has an action

Fig. 3-2. Compression-type safety released by a gear.

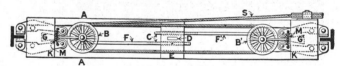

Fig. 3-3. Compression-type safety applied by a rope and pulleys.

similar to that in Fig. 3-1, except that rope and pulleys are used instead of drum and drawbars to tighten the safety shoes onto the guide rails. Tail rope S is reeved through pulleys B and B', each with five sheaves. Figure 3-5 shows the governor roping arrangement for the safety in Fig. 3-3. Instead of the governor rope being endless as in Fig. 3-4, one end is dead-ended to the car at A. From here the rope passes down around the weighted sheave in the pit up over the governor sheave and down under the car around the pulley sheaves to another dead end (Fig. 3-3). The governor rope in Fig. 3-5 is free to move with the car and drive the governor sheave, as is the one in Fig. 3-4.

When the governor trips and grips the rope the down motion of the car pulls the rope through the sheaves to draw them together. This action pulls wedges G and G' between the inner ends of the

safety-shoe levers.   From here on the action of this safety is similar
to that shown in Fig. 3-1.   Reset rods *F* and *F'* connect to the outer
ends of the pulley yokes, but the inner ends are free to slide in
grooves in casting *E* and have notches *C* on their upper surface.

To release this safety on a drum machine, after the hoist ropes are
in place on the drum and taking the weight of the car, the governor

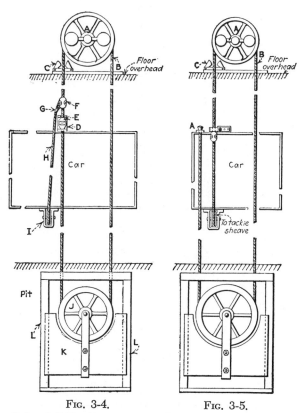

FIG. 3-4.	FIG. 3-5.

FIGS. 3-4 and 3-5.   Arrangements of two safety-governor ropings.

rope is released from its grip.   A special flat steel bar wrench is inserted through an oblong opening in the car's floor, the end of the
bar projecting through the opening down between bars *C*.   When
the bar is rocked, dogs on either side of it engage teeth in bars *C*
and force them with the pulleys in opposite directions.   Repetition
of this operation forces the wedges from between the inner ends of
the safety-shoe levers and spring *M* returns these levers to their off
position and lifts the safety shoes clear of the guide rails.

Up to now the discussion of safeties has been more or less diagrammatic, but in Fig. 3-6 the equipment is shown for a passenger car, supported in a sling built of steel channels. At the top of the sling are two guide shoes T and at the crosshead center are attached

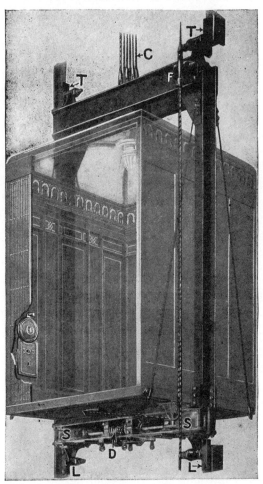

Fig. 3-6. Car with its compression-type safeties supported in a sling made of steel channels.

the hoist ropes C. At the bottom of the sling, on which the car rests, is the safety plank in which the safeties are located. Jaws that grip the guide rails are at S and the lower guide shoes at L. The tail-rope drum for applying the safeties is at D, with the governor-rope releasing carrier at F. These parts can be identified on

Figs. 3-1 and 3-4, in connection with which their operation has been explained.

Vertical-shaft Flyball Governors. Elevator governors operate on principles similar to those of other mechanical speed-controlling devices in which weights, flyballs, arms, and links are held by spring tension, gravity, or both against centrifugal action produced by the balls rotating about an axis. There are several types of governors, some vertical-shaft and others horizontal-shaft, but they all operate on the same principle. In the governor shown in Figs. 3-7 and 3-9 sheave *A* is driven by a rope which in turn drives the flyballs *O* through bevel gears at *B*. As the balls are thrown outward, through

Fig. 3-7. One design of vertical-shaft flyball governor.

Fig. 3-8. Horizontal-shaft governor with weights in the driven sheave.

connecting links *E*, they pull sleeve *C* upward against the tension of spiral spring *D*.

Sleeve *C* carries with it divided lever *F* pivoted at *G*. Upward motion of the lever transmitted through adjustable connection rod *H* lifts lever *I* to trip the governor-rope gripping jaws *M* and *N*. Once unlatched the jaws fall in contact with the downward moving rope to clamp it tightly, stopping further motion and setting the car safeties. Jaw *N* is supported in slotted holes in the frame against the pressure of spring *S*, which gives the jaws flexibility to ease their clamping action on the governor rope.

Horizontal-shaft Governors. Figures 3-8 and 3-10 show a horizontal-shaft governor in which two flat curved weights *B* and *B* are pivoted a short distance from their small ends to the sheave spokes at *C* and *C*. Movement of the weights is synchronized by rod *D*

(Fig. 3-10), and their outward motion is opposed by spring *E* on eyebolt *F* pivoted at *G* on one of the weights. Shank of the eyebolt passes loosely through lug *H* on a sheave spoke and has an adjustment that is sealed when the governor is adjusted to set the safeties at a given car speed. When the set overspeed is reached weights *B* move out and strike the heavy U-shaped arm *F*, carrying it upward toward the vertical center line. This arm is pivoted eccentrically at *C* in Fig. 3-8, so that its extended jaw moves more deeply into the sheave's groove as the arm approaches the vertical. Motion of the governor rope completes the clamping action as soon as contact is made with the rope to set the car safeties.

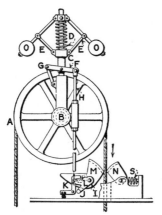

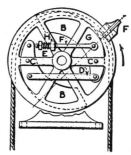

Fig. 3-9. Diagram of the governor shown in Fig. 3-7.

Fig. 3-10. Diagram of the governor shown in Fig. 3-8.

Governor Switches. On high-speed elevators many of the governors have an overspeed actuate switch in the control circuits to cut power out of the motor and stop the elevator. On all new elevator installations for car speeds above 150 fpm The American Standard Safety Code for Elevators requires that all overspeed governors shall have a switch that opens when the car is in down motion at a speed of 90 to 95 per cent of that at which the governor is set to trip, depending on the rated car speed. Figure 3-11 shows a governor with a double-contact switch, at its lower left-hand side.

Inertia-type Governors. On high-speed elevators combination speed- and inertia-type governors have been used. Normally the governor functions as a flyball type. If for any reason the car accelerates at a high rate, as when falling, the inertia element comes into effect and sets the car safeties, when it is moving at a comparatively low speed. The inertia element is a weight that at normal speed accelerates with the governor. At high rates of acceleration, weight movement lags behind that of the governor and causes the safeties to be applied to the car.

It might seem that the overspeed governor on an elevator is a rather unimportant part of the equipment. It performs no service in normal operation of the elevator and it may not be called upon to act during the life of the machine. However, if the other safety

Fig. 3-11. Vertical-shaft flyball governor with control switch.

devices, such as the car switch, emergency switch, mechanical brake, or dynamic brake, fail to prevent excess speed, or if the hoist ropes break, then the governor is called on, first to open the potential-switch magnet-coil circuit and then to apply the car safeties.

It is therefore evident that reliability is of first importance in governor operation. Its design must be such that wear or rust will not prevent operation at a critical time. It must be sensitive to overspeed yet stable, so that it will not operate accidentally. Normal speed should cause slight movement of the flyballs. This indicates that they are free to move.

Releasing Carriers. The governor is driven through its rope from the releasing carrier bolted to the top of the car or to the crosshead, as shown at *F* in Fig. 3-6. This carrier overhangs the edge of the car and holds in a flexible grip that leg of the governor rope which applies the gripping jaws. The carrier releases the governor rope when its motion is stopped by the tripping of the governor. Then, the only connection between the governor rope and the car is through the tail rope *H* shown in Fig. 3-4, which leads to the safety

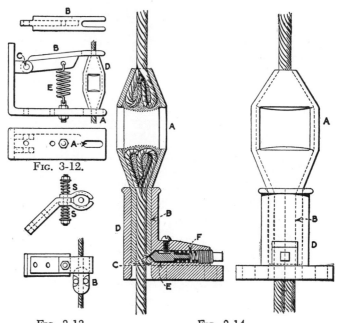

Fig. 3-12.

Fig. 3-13.   Fig. 3-14.

Figs. 3-12 to 3-14.   Types of governor-rope releasing carriers.

equipment under the car and which is actuated by the downward motion of the car.

Figure 3-12 shows one type of governor-rope releasing carrier. The governor rope passes through an elongated hole *A* in the cast-iron base and is held by divided finger *B*, pivoted at *C*, holding the double-ended connector *D*, aided by spring *E*. When the governor trips and grips its rope, downward motion of the car pulls the rope out of the slot in arm *B* to set the car safeties.

Another type of releasing carrier (Fig. 3-14) uses a modified double-ended rope connector *A* with an extended flattened shank *B* having a notch *C*. This is held in a cast-iron base *D* by a pointed trigger *E* and a spiral spring *F*. As shown on the right of Fig. 3-14, both the governor and the safety tail rope are socketed in the lower basket of connector *D*.   In the releasing carrier shown in Fig. 3-13, stop ball *B* is held beneath two curved jaws normally held closed by two spiral springs *S*. When the governor trips, the releasing jaws open from the increased pull on the rope and the stop ball slips through to free the rope and set the safeties.

Tension in Governor Rope to Apply Safeties.   The rope grip on the governor must stop the governor rope so that further descent of the car will cause the rope to be pulled loose from the releasing carrier on the car and turn the safety drum.   This action will engage the safety device, after which descent of the car increases pressure of the safety jaws against the guide rails until the car stops.   It is possible that this force may become so great that some part of the

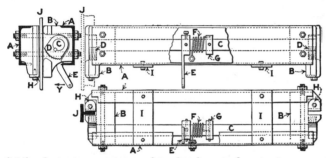

FIG. 3-15.   Instantaneous-type safety that has single gripping eccentrics.

safeties or the governor ropes would be overstrained before the car stops.

To eliminate this danger the governor rope grip has a spring, as shown in Fig. 3-9, or some other device that permits the rope to slip, after sufficient force has been applied to stop the car, but before overstressing the safety mechanism.   The governor rope in any case must be strong enough to pull loose from the releasing carrier.   Tension necessary to slip the rope in the governor's gripping jaws varies with the type of safeties and the elevator capacity.

Single-eccentric Safeties.   Car safeties may be of the gradual or compression type, already described, for car speeds over 125 fpm, or instantaneous or clamp type for car speeds of not more than 125 fpm.   Figure 3-15 shows a single-eccentric safety of the latter type for light-duty cars.   Channel irons *A* bolt to the safety heads *B* by through bolts.   These heads are heavy castings that have renewable brass wearing plates or gibs *H*, making them combination safety heads and guide shoes. On each end of rocker shaft C are keyed gripping eccentrics D. Tripping lever E connects to the governor rope to apply the safeties in case of car overspeed. Torsion spring F normally holds the rocker shaft and its eccentrics in release position and returns the safeties to this position after being freed when they have set by overspeed. One end of spring F connects to collar G pinned to the rocker shaft and the other end bolts to one of the channels. Plates I stop the car on the bumpers if it runs by the bottom landing, and J are the guide rails.

When the governor trips from overspeed and grips the rope so that it cannot travel with the car, the rope is pulled free of the releasing carrier. Further down movement of the car causes lever E

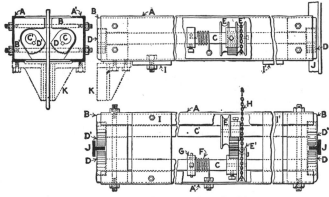

Fig. 3-16.　Instantaneous-type safety that has double gripping eccentrics.

to turn the safety eccentrics toward the guide rails. When their corrugated surfaces contact the rails, the eccentrics are carried the remaining distance by the car to lock the safeties to the guide rails and stop further motion. Instead of using eccentrics it is now common practice to lift a knurled roller up between the safety jaw and guide rail.

Double-eccentric Safeties. Figure 3-16 shows a heavy-duty instantaneous safety. It has two rocker shafts C and C' with eccentrics D and D', instead of one. Segmental gear sheaves E and E' keyed to the two shafts synchronize their motion. A tripping chain H that runs in a groove in one of the sheaves sets the safeties when the governor trips from overspeed. This action is similar to that shown in Fig. 3-15, except that the rails are gripped by double eccentrics instead of between single eccentrics and gibs.

To release the car after the safeties shown either in Fig. 3-15 or in Fig. 3-16 have set, it is only necessary to wind the hoisting ropes back on the drum, taking up the slack slowly. Then, the car should be raised by holding the brake open and using a spanner wrench or bar, usually provided for the purpose, on the brake-wheel coupling to turn the worm gear and drum.

Lifting the car may also be done by operating the controller by hand, first blocking open the accelerating switches. But this method is likely to be abrupt, imposing a severe strain on the ropes. As the car lifts, the eccentrics roll downward until free of the guide rails, when they are snapped into normal position by the torsion springs. After the governor-rope gripping jaws have been released, the rope is restored to its anchorage in the releasing carrier. Cause of the

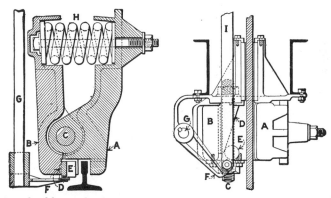

Fig. 3-17.   Flexible guide-clamp-type safety has heavy springs to ease its application.

safeties setting, if other than the hoisting ropes parting, must be found and corrected. Also, the guide rails must be examined to determine if they have been roughened, displaced, or otherwise damaged.

Flexible Guide-clamp Safeties. Figure 3-17 diagrams a flexible guide-clamp safety that has two heavy jaws A and B pivoted at C. Jaw B has a wedge D with a loose cylindrical serrated roller E, normally held in position at the lower or thin end of the wedge as shown. When the governor trips and lever F swings by action of the tail rope on rocker shaft G, the roller is pushed up until it contacts the guide rail. Motion of the car then continues to move the roller up between the rail and the inclined surface of the wedge. Jaw A pulls against the rail, and as the roller moves up heavy spring H compresses until pressure between jaw A and the rail becomes great enough to stop the car. The springs ease the application of the safeties, and the car is not stopped with the shock that occurs with some other types of instantaneous safeties.

Counterweight Safeties. Where space below the hoistway is used as a passageway or is occupied by persons, counterweights are required to have governor-actuated safeties, similar to those on the car. Governors for counterweight safeties should be set to trip at a speed greater than, but not more than 10 per cent above, that at which the car governor is set to trip.

Car and counterweight safeties are classified in The 1955 American Standard Safety Code for Elevators as types A, B, and C. Type A safeties are instantaneous designs (Figs. 3-15 and 3-16) that apply pressure to the guide rails through eccentrics, rollers, or similar devices. Due to the design of the safeties they stop the cars quickly and are suitable for low-speed cars only, below 125 fpm.

Type B safeties are wedge-clamp designs (Figs. 3-1 and 3-6) that apply a gradually increasing pressure on the guide rails during the stopping period. These, when correctly designed, are applicable to cars of any speed and are required for rated car speeds above 125 fpm. A type C safety is a type A with oil buffers. It develops a retarding force on the car during the compression stroke of one or more oil buffers. It is limited to cars operating at 500 fpm or less.

On some cars two safeties have been used in multiple. Duplex or dual safeties can be only type B. Types A and C are not permitted to be operated in multiple.

Elevator car safeties and governors have been developed in many forms, but the foregoing gives a general picture of the principles used. For more details of their construction see American Standard Practice for the Inspection of Elevators, Inspectors' Manual.

CHAPTER 4

Brakes, Their Care and Adjustment

Mechanically Operated Brakes. On the older designs of elevator machines, the brake is one of the most important parts of the equipment, for upon its proper functioning depends not only the safety but also the satisfactory operation of the machine. If the brake fails to function, control of the car may be lost by the operator, with serious results. If the brake is applied too harshly, the car will be stopped with an unpleasant jerk to the passengers and the equipment will be strained unnecessarily. When the brake is not applied with sufficient force, the car will be difficult to control and it will be hard to make floor landings. On modern d-c elevators the mechanical brake is used primarily for holding the elevator at rest after it has been stopped by electrical braking.

Elevator brakes have been made in about as many forms as there are types and makes of machines, so that about all that can be done here is to indicate a few of the different types and how they operate and are adjusted. Figure 4-1 diagrams an earlier type of elevator-machine mechanically operated brake. The two cast-iron shoes S and S' are hinged at H and attach to lever L at A. Lever L connects to a second lever B that is raised and lowered by a cam C on the shipper wheel, the latter not shown. Rotating the shipper wheel to either running position by the operator in the car pulling the hand rope lifts the left-hand end of lever B, and with it lever L. This allows shoe S to drop away from the brake wheel until it is brought to rest on stop D, after which shoe S' is lifted by a further upward movement of lever L. Application of the brake is the reverse of this process, and the force that applies the brake is controlled by the position of weight W.

In general a brake should be adjusted so that its shoes just clear the wheel. However, where a brake is operated mechanically, it may be necessary to adjust it to open considerably more than this so that it will not be applied before power is cut out of the motor. When the brake is adjusted (Fig. 4-1), screw *D* is put in a position that will let brake shoe *S* drop clear of the wheel, and screw *E* is then adjusted so that shoe *S′* will clear the wheel by the same amount as shoe *S* when the operating mechanism is in full-on position. After this has been done, the operating mechanism should then be pulled slowly to off position and a check made on the controller to see that the brake cannot be applied hard while the motor is still connected to the line. If such a condition exists, it will be necessary to give both brake shoes more clearance from the wheel. After the brake shoes have been properly adjusted, weight *W* is

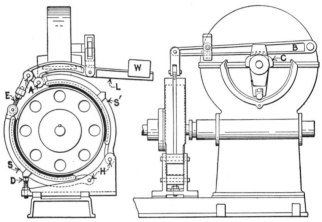

Fig. 4-1.   Mechanically operated brake.

moved to a position that will stop the car within the desired distance.

On some older types of machines the starting resistance could be cut into circuit before power was cut out of the motor. Then, if the brake were applied before the motor was disconnected from the line, it would be possible to stop the motor with the starting resistance in series and connected to the power supply, in which case the starting resistance would overheat. In one case in mind, the operator left the elevator stopped this way when he left the building on Saturday noon. When he returned Monday morning everything combustible on the motor and controller was burned up, even the wood terminal blocks. Probably the only thing that saved the building was that the equipment was installed in a fireproof room. Such occurrences were not uncommon with some of the old types of mechanically operated machines.

If the brake can be applied before power is cut out of the motor, fuses are likely to blow, if resistance is not cut into the motor circuit. If fuses do not blow, breaking the heavy current on the controller contacts will cause serious burning on these parts. Improper brake adjustment, especially on some of the older types of machine, was not infrequently the cause of considerable controller trouble.

Instead of using cast-iron shoes lined with leather as in Fig. 4-1, some older types of brakes have a steel band lined with leather. The two ends of the band connected to the brake lever, as shown in Fig. 4-1, with the band center made fast so that the brake can open from both ends. One difficulty with a band brake is to get it to clear the wheel at all points when released, since the band does not bend to a true circle about the wheel.

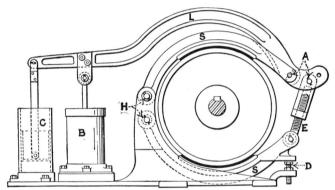

Fig. 4-2.   The brake shown in Fig. 4-1, modified for electrical operation.

Electrically Operated Brakes. Figure 4-2 shows a somewhat similar type of brake to that in Fig. 4-1, except that it is solenoid operated. The shoes, adjusting screws, and lever are lettered as in Fig. 4-1. Lever *L* is brought back over the brake wheel and is connected to solenoid *B* and an air dashpot *C*.

The solenoid is energized from contacts on the controller. Energizing coil *B* causes it to pull up its core and lift the lever, thus clearing the brake shoes from the wheel. Dashpot *C* controls how quickly the brake is applied. After the brake shoes have been adjusted so that both just clear the wheel when solenoid *B* has pulled up its core, the dashpot should be adjusted to allow the brake to be applied easily. The correct adjustment of the dashpot can be determined by starting and stopping the car. On this type of brake, if the shoes are given too much clearance from the wheel, they may allow the lever to drop so far that the dashpot will bottom before the brake is applied, in which case it will be difficult to stop the car. It may also be possible that the core in the solenoid will drop down so far that the coil cannot develop sufficient pull to release the brake. For this reason care should be exercised to get the brake properly adjusted. With a magnet-operated brake, contacts on the controller that make and break the coil circuit must be kept adjusted so that they will always break the circuit when power is cut out of the motor and make the circuit when the motor is connected.

Equalizing Brake-shoe Clearance. On the two brakes discussed the lever has no fixed fulcrum. Therefore brake shoes cannot be

Fɪɢ. 4-3.   Electrically operated brake applied by spiral springs.

adjusted so that one applies harder than the other, although one shoe may be released or applied ahead of the other. With the type of brake shown in Fig. 4-3, brake lever *H* has a fixed fulcrum pin *E*, about which the lever moves up and down. In this brake, leather-lined shoes *S* are applied by springs *J* instead of by a weight. Each shoe is attached to lever *H* by turnbuckles *K* and *K'*, and the long arm of the lever is attached to core *G* of solenoid *F*. When coil *F* is energized, it pulls up its core and the brake lever. In so doing, it pushes turnbuckle *K* up, and with it the top brake shoe against the tension of spring *J*. Likewise, turnbuckle *K'* and the bottom brake shoe are pushed down against the tension of spring *J*, thus lifting the shoes clear of the wheel. When the coil circuit is opened, the core and lever are released and the brake shoes are applied by the springs, assisted by the weight of the solenoid core on the end of the lever.

In adjusting the type of brake shown in Fig. 4-3, it is essential that both shoes clear the wheel equally so that they will be applied with equal force to the wheel. If turnbuckle K' is made considerably longer than K, it may be possible to have the top shoe applied hard on the wheel and the bottom shoes not touching. After the top shoe has set on the wheel, turnbuckle K would then hold the lower shoe away from the wheel. It is easy enough to adjust the shoes to the same clearance, simply by blocking the end of the lever up into the release position and then turning the turnbuckles to obtain the desired clearance for each shoe.

Equal clearance of the brake shoes does not necessarily give the best stopping of the car, although it is the best starting point in the final adjustment. The type of brake, amount of counterweight, and average loading of the car, and dynamic-braking action of the motor all have an influence. With the type of brake shown in Fig. 4-3, where one shoe can be applied harder than the other, it may be found that the car will stop more quickly in one direction than in the other. In extreme cases, the car may stop with a decided jar in one direction and slide in the other. This is an indication that one brake shoe is being applied harder than the other; which one can be determined by inspection or a little experimenting. First try setting one shoe a little closer to the wheel and operate the car again. If this adjustment improves the setting, it is known that the adjustment is being made in the proper direction. If the stopping is not improved or is made worse, the wrong shoe is being adjusted. After the brake has once been properly adjusted, any adjustment for wear should be made equal on each shoe by marking the turnbuckles and turning each an equal amount.

Double-core Magnet Brakes. The brake shown in Fig. 4-4 is one that has a small movement of the various parts during operation. A single magnet coil E with a double core D and D' is used for releasing the brake, and the shoes are applied to the wheel with spiral springs S. The brake levers are fulcrumed at F and the shoes are attached at A. When coil E is energized, it pulls the two cores D and D' together and in so doing lifts the shoes off the brake wheel. Since the shoes are hinged at their middle and lifted away from the wheel at its horizontal diameter by the levers, they are lifted an equal distance from the wheel at all points, which is not true of some brakes that have their shoes hinged at one end. Lifting the shoes equally at all points on their periphery allows for minimum movement when releasing and applying the brake.

The movement of cores D is adjusted on the screws E until, when the inner ends of the cores are in contact, the brake shoes just clear the wheel. Then screws H are adjusted to ensure equal movement of each brake shoe. These are adjusted so that they will be about touching stops at H' when the brake is released and the shoes are an equal distance from the wheel. Pressure that the brake applies to

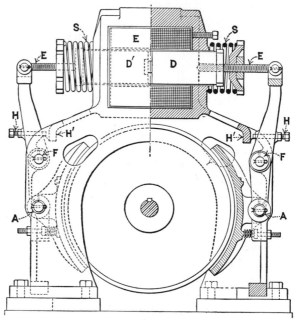

Fig. 4-4.   Brake used on one of the earlier types of medium-speed elevators.

the wheel is obtained by adjusting spiral springs S. This tension should be such as to stop the car without undue jar to the passengers. In this brake one shoe is applied independently of the other; therefore the springs should have equal tension. This can be determined with sufficient accuracy by measuring the length of each spring; they should be of equal length.

As previously mentioned, adjustments are made by screwing stems E in and out of the cores D and D', so that when the cores are pulled together the brake shoes are lifted clear of the wheel. As the lining of the shoes wears, stems E should be backed out of cores D and D' periodically to maintain this adjustment. If such an adjustment is

retained until the shoe linings are worn to where they must be renewed, when the linings are put into the shoes, core D and D' will
probably touch when the brake shoes are in contact with the wheel.
To correct this condition, screw the stems into the cores to separate
the latter enough to give sufficient movement to lift the shoes clear
of the wheel. In adjusting the brake, care must be exercised not to
separate the cores too much, or the coil may not develop sufficient
pull to lift the shoes from the wheel.

Mechanically and Electrically Operated Brakes. The brake shown
in Fig. 4-5 is mechanically operated, but is applied electrically if the

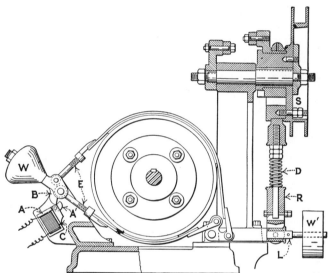

Fig. 4-5. Combination mechanically and electrically operated brake.

power supply fails. Brake lever L connects to shipper wheel S by
adjustable rod R. Coil C connects directly across the power line.
As long as the line is alive, coil C is energized and holds armature A
in the position shown. Extension A' is part of armature A, and B
is an extension of the lever carrying weight W. As long as armature
A is held to the coil, weight W will be held in the position shown.

Under normal operation the brake is released and applied by the
movement of shipper wheel S raising and lowering lever L. Tension
with which the brake is applied is controlled by spring D, as well as
by the position of weight W' and adjustment of screws E. If power
fails, coil C releases armature A and weight W will fall and apply
the brake, which cannot be released by the mechanically operating

mechanism until weight W is raised and held in its off position. This prevents the operator from leaving the brake released on the machine with power off the motor. On this type of brake, one brake shoe can be applied harder than the other if not properly adjusted. Therefore careful attention should be given to this feature when making adjustments.

Care of Brakes. Before a brake can be expected to work properly the surface of its linings must be in good condition. There are three common materials used to line brake shoes—namely, asbestos, leather, and wood—and preference for these materials is about in the order named. Keep lining surfaces free of oil and dirt. Oil may work out along a worn shaft from the gear case and get on the brake wheel. When this occurs, remove the shoes and scrape off the oil and dirt with glass or wash off with gasoline. Fuller's earth may be used to absorb the oil, after which it should be scraped off to clean 'the lining surfaces. Leather brake linings sometimes become dry and hard, in which case they should be treated with a good leather dressing or neat's-foot oil.

When a brake is adjusted, the electrical features cannot be overlooked. In most modern d-c elevator controllers, the armature of the motor, after being disconnected from the line, is connected through a resistance. This causes the motor to act as a generator and offer a retarding effect on the motion of the machine, known as dynamic braking. To ensure satisfactory stopping the resistance of the circuit must be a correct value, but this is generally properly adjusted when the installation is put into service and requires no further attention except in case of short circuits or open circuits in the dynamic-braking resistance. On high-speed elevators, connection schemes are used in the brake circuit to accommodate the braking action to the different speeds and loadings of the car at the time of stopping.

Brakes on High-speed Elevators. Conditions under which an elevator must be stopped range from full load in up motion to full load down. Figure 4-6 shows a type of brake used on one make of medium- and high-speed direct-traction machines. It operates through a dual-lever system. Brake-shoe levers O are fulcrumed at Q and have tension springs I acting on their top ends to apply the shoes to the brake wheel. Top levers M are fulcrumed near their lower ends, connect to magnet cores A, and at their top ends connect to switch N. When the brake coil is energized cores A are pulled together to move the lower ends of levers M outward. These push the top ends of levers O apart to release the brake.

Figure 4-7 shows the electrical circuits that operate the brake shown in Fig. 4-6. There are four resistors in these circuits: (1) *BR*, brake; (2) *ABR*, auxiliary brake; (3) *PBR*, parallel brake; and (4) *PBR'*, auxiliary parallel brake. Resistance *ABR* reduces the current in the brake coil after it has lifted the brake shoes clear of the

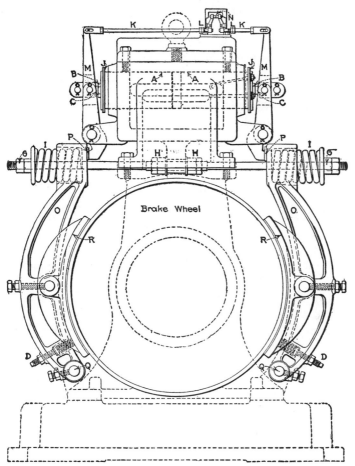

Fɪɢ. 4-6.   Diagram of one type of brake used on medium- and high-speed elevators.

wheel. This reduces heating of the coil and permits quick release of the brake at starting. Resistances *PBR* and *PBR'* provide a discharge current circuit for the brake coil when its exciting circuit is opened. This current prevents the magnetic field of the coil from decaying rapidly, thus slowing down application of the brake.

Brake-coil circuits provide three different applications of the brake. At terminal landings a more positive brake action is required then at intermediate floors. Stopping at intermediate floors must be more gradual because, when running at high speed, if the brake should be applied immediately when making a stop the car would be given an abrupt and unpleasant jolt. The third application of the brake is with car-switch control when inching to the floors to level the car platform with them. This requires close adjustment of the brake shoes to the pulley and a quick-acting brake. If the brake shoes' lift is too great, time required for the shoes to move through the clearance will permit the car and load to overhaul. Adjustment of the brake to obtain best operation will now be considered.

Adjusting Brakes on High-speed Elevators. 1. Brake shoes and pulley must be perfectly clean.

2. Magnet cores A and A are adjusted (Fig. 4-6) to obtain the minimum lift by unlocking nuts B and B and moving cores A and A

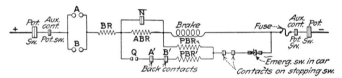

Fig. 4-7. Diagram of the electrical circuits for the brake shown in Fig. 4-6.

either in or out as desired. Magnet cores A and A are drilled and tapped, and are moved in or out by turning the cores on the threaded studs C and C. The lift at the top and bottom of each shoe is equalized by screwing in or out studs D and D to obtain the required results.

3. Nuts L and L are adjusted on braking switch N so that when the brake magnet cores have pulled in, the contacts are opened the minimum distance required to break the arc.

4. Nuts G and G are tightened to obtain the spring tension necessary to give the desired stop with full load. The nuts H and H should be slacked off while this is being done and then should be pulled up equally to ensure equalizing tension on springs I and I.

5. Leather washers J and J keep cores A and A from butting and should be of sufficient thickness to provide a positive release of the cores when power is cut off. These washers provide for a softer blow when the magnet cores pull in by not striking metal to metal.

A-C Brakes. In a-c elevator work the design of a satisfactory brake is quite a different problem from that of a d-c one. Stopping of an elevator operated by a d-c motor is done by the use of two brakes, the dynamic and the mechanical. Dynamic brake action is produced during the stopping period by connecting the motor's armature to a resistance, so that it becomes a generator and offers a retarding force to the elevator's motion. In this way, energy stored in the moving mass is dissipated in the form of heat in the resistance. When motor speed is high, effect of the dynamic brake is strong, but this effect decreases as the machine slows down and becomes zero when the motor stops. Under these conditions the mechanical brake has very little work to do, its chief function being to hold the car after it has come to rest; therefore the resulting action of the two brakes can be made very smooth.

An a-c motor cannot easily be used as a dynamic brake, particularly at low speed. Consequently the mechanical brake must be of sufficient capacity to dissipate all energy from the motion of the machine parts and load. With single-speed a-c motors the mechanical brake must stop the car from full speed. If multispeed motors are used, the motor can be utilized to slow the car down to a speed corresponding to the slow speed of the motor, but below this the mechanical brake must stop the car. Consequently, an a-c elevator brake must be considerably larger than that for a d-c machine of a corresponding capacity and speed. An a-c motor of the same capacity and speed will have a considerably heavier rotating member than a d-c machine. This, taken in conjunction with the larger brake wheel, adds to the energy that the brake on an a-c elevator must dissipate, but, as stated in Chap. 8, a-c brakes are becoming obsolete.

Work Done by a Brake. Evidence of the amount of work done by a brake is the temperature of the brake shoes and brake pulley. Brake-shoe linings that last many years in d-c elevator service will last only a few months on a-c elevators. In a-c elevator service, the heavy brake spring pressure required produces an objectionable effect that is difficult to eliminate, that is, the abrupt stop so characteristic in this class of elevator service.

Brake Magnets. A-c brake magnets have characteristics that are quite different from those of d-c magnets of similar design. A d-c magnet takes the same current regardless of the position of its armature, but the pull of the magnet increases rapidly as the air gap is reduced. An a-c magnet takes several times as much current when the air gap is large as it does when the magnet is closed. Most a-c magnet coils will develop an excessive temperature and burn out if left with current on and the armature open. The pull of an a-c magnet is more nearly constant throughout its stroke than that of the d-c magnet.   A d-c magnet pulls more slowly and the action is smoother and softer than in the a-c type.   The action of the a-c magnet is more abrupt both in closing and in release.   This is one of the reasons oil dashpots are used on some a-c brakes.

In an a-c magnet, on account of the alternating magnetic field produced by the current, magnetic circuits are made of laminated iron or steel, instead of being solid as in d-c work.   If a solid core were used for a-c work, eddy-current losses in it would be so high as to cause excessive heating and in other ways interfere with the satisfactory operation of the magnet.   If single-phase magnets are used, a short-circuited copper loop (generally known as a shading coil) must be used in the face of the magnet core to reduce the noise when in operation.   The difficulties encountered in the use of single-phase a-c brake magnets have caused some elevator manufacturers to use polyphase magnets or d-c brakes powered from a metallic disk rectifier (see Chap. 8).   Some of the a-c designs are of the plunger type, and others are of the rotating magnet type.

Polyphase Magnet Brakes.   Figure 4-8 shows a cross section through a polyphase a-c brake.   In this design a slotted laminated core supports four coils, A, B, C, and D.   To obtain a more even distribution of the pull over the face of the magnet the coils overlap each other. On the left-hand side of the magnet, counting from the top, coil A is in slots 1 and 3, while coil B is in slots 2 and 4.   Coils A and C are connected in series and to one phase, whereas coils B and D are connected in series to another phase of the power supply.   Armatures E are supported in loops F, which in turn are supported at the top and bottom, and act as guides for the movement of the armatures. Plungers P are also attached to loops F and act as pistons in the dashpots G.   This whole structure is made up as a unit and is supported from the cover of the tank in which it is immersed in oil.

Connection is made between brake levers H and dashpot piston P by gooseneck links N.   It might appear that the gooseneck links would result in a binding or twisting action.   This, however, is not the case, since the connection to the pistons on the inside of the oil pot is directly in line with the fulcrum-pin connection to the brake lever on the outside, and gives the same effect as a straight rod brought out through the side of the oil pot to the brake lever.

The brake-shoe levers are fulcrumed at J, and springs S, which are in compression, force the hardwood-lined brake shoes K onto brake wheel W.   When the brake is adjusted, stroke-adjusting studs R are screwed into or out of the gooseneck links N until, when the magnet is closed, the shoes will just clear the brake wheel.   Capscrews *M* are screwed in until they just about touch the shoes. These are intended to allow the shoes free play to adjust themselves to the wheel and at the same time to prevent their top ends from falling over and riding on the wheel when the brake is released. When the brake has been adjusted to obtain the proper clearance, the force with which the shoes are applied to the wheel is adjusted by changing the tension of springs *S* with adjusting studs *O*. The same amount

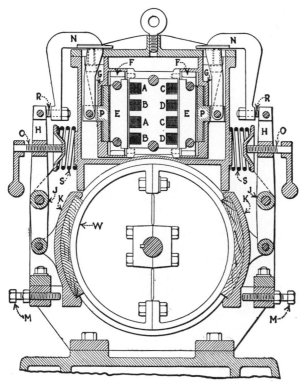

Fig. 4-8.   Cross section through the polyphase brake shown in Fig. 4-9.

of tension should be given both springs and can be determined by their length, which should be the same for each. These springs can also be seen on the complete assembly of the brake in Fig. 4-9.

When the coils are energized, armatures *E* are attracted to the core, which moves the top ends of levers *H* toward the magnet and the brake shoes away from the wheel, and compresses springs *S*. De-energizing the coils releases armatures *E*, and the springs push the top ends of levers *H* away from the magnet and the shoes onto the wheel. The movement of the armatures toward the magnet also moves pistons *P* out of the dashpots and tends to create a vacuum, thus temporarily reducing the effective pull of the magnet. When the magnet is deenergized, the oil that has flowed into the dashpot is compressed and the application of the brake shoes is retarded. Thus, the same dashpot provides a retarding force in either direction.

Each brake is tested and adjusted for rapidity of the stroke at the factory, and no attempt should be made after installation to alter the dashpot action unless abnormal conditions make this necessary.

Fig. 4-9. Polyphase magnet immersed in oil mounted above the brake wheel.

Plugs are provided in the dashpots, which can be replaced by others having suitable openings for the possible adjustment of the action in case of necessity.

It is important that the oil be maintained at the proper level in the brake magnet housing. If the dashpots are not completely covered with oil, air will enter, causing the brake to slam and oil to be sprayed out of the case. If oil gets on the brake shoes and wheel, the coefficient of friction will be reduced and the car will slide. The oil should be scraped off the shoes and washed off the pulley with gasoline. It should be changed about every six months, because oil sludges when subjected to heat and exposed to the atmosphere. The manufacturers use a special grade and substitutes should be avoided.

Nonsealing Reciprocating-type Brake Magnet. The brake magnet shown in Fig. 4-10 is nonsealing and is of the polyphase rotating-magnetic-field type, but reciprocating in operation instead of rotating.

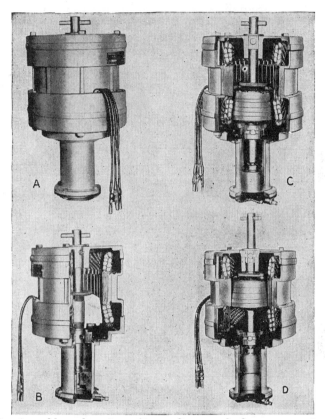

Fig. 4-10.   Assembly and sections of nonsealing-type brake magnet.   *A*, assembly of brake magnet.   *B*, section through stator and movable core.   *C*, section through stator showing movable core in released position.   *D*, section through stator showing the movable core in full-stroke position.

As seen in the figure, the stationary part is similar to a polyphase induction motor's stator with a laminated core and a distributed winding. The rotor has a laminated core without a winding displaced out of line with the stator when the magnet is in the open position, as shown in *C*. There is no rotative effort but a tendency for the core to align itself with the stator when the latter winding is energized. It will be seen that since the rotor core tends to align itself with the stator core, it does not require a mechanical stop and therefore cannot cause a slapping noise. Because a polyphase distributed winding on a magnet produces a pull that is always considerably above zero, it will not cause a magnetic hum or vibration. By shaping the core's face it is possible to produce practically any characteristic stroke-pull curve desired; as the constant pull is most desirable for brake service, this is used so as to eliminate any violent action of the magnet. This type of brake magnet is used on the elevator machine shown in Fig. 1-44.

There are many other makes and designs of a-c elevator brakes, but those described are representative of the general principles involved. The brake mechanisms in general are not radically different from the d-c type, the chief differences in most cases being in the magnet's construction, the use of dashpots, and, in some designs, immersing of the magnets in oil.

CHAPTER 5

Direct-current Semimagnetic Controllers

Types of Controllers. Elevator controllers may be divided into two general classes, semimagnetic and full-magnetic. The semimagnetic types are those used on elevators that are operated by a hand rope running through the car, and are now fast going out of use. The operator may control the elevator either by pulling on the rope directly or by a lever or wheel. On the semimagnetic control, the reversing switch is generally operated directly by the operator in the car, through the operating rope, after which the starting resistance is cut out of circuit automatically by a magnet or solenoid.

Sometimes the semimagnetic control is referred to as a mechanical type. This, however, is not correct, since it would indicate that all the functions of the controller are performed mechanically. In most cases in which elevators are controlled with a hand rope, part of the control functions are performed by a magnet or solenoid; hence the name semimagnetic control.

Some of the earlier types of elevator controllers were constructed so that all their functions were performed mechanically, no magnets being used in their operation. Such control systems may be classed as a mechanical type, but today they are seldom found in use.

Broadly, full-magnetic controllers are those in which all the functions of the control are performed by magnets. The initial impulse for starting and stopping the car may be from a car switch or push button, after which the controller completes the operation automatically. These types of controllers may be divided into car-switch, automatic leveling, automatic landing, push button, signal control, dual control, etc., depending upon what the control equipment accomplishes.

Compound Motors. Semimagnetic controllers are generally designed for but one speed, that is, the car can be run at full speed only, no slowdown being provided. Consequently the equipment is

111

only applicable to slow-speed machines. Compound motors are usually used, with the armature, shunt-field coil, and series-field coil leads brought out to separate terminals.   When the motor is running full speed, it is a shunt machine, the series-field coils being cut out,

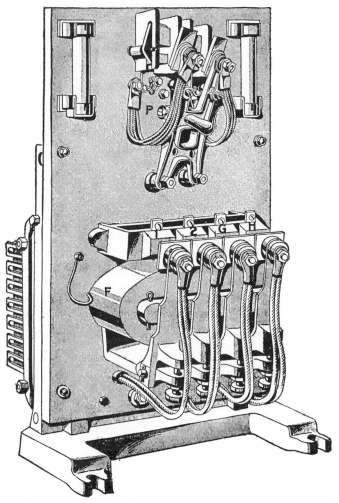

Fig. 5-1.   Elevator-motor controller with an accelerating magnet *F* and a potential switch *P*.

through the accelerating magnet, as this gives better speed regulation under various loads.

Single-speed Controllers.   The controller shown in Fig. 5-1, which is an early type used by Otis, consists of a main-line or potential switch *P*, which is open when the machine is at rest and is closed by a magnet coil through contactors operated by a cam, located at *A* in Fig. 5-2. The potential switch has a magnetic blowout coil to prevent the arc from being carried between contacts when the switch opens. The accelerating magnet *F* is a coil of fine wire fitted over an oval-shaped core on which four contact arms are arranged, two of which (1 and 2) cut out sections of the starting resistance, the other two (*G* and *H*) cutting out the series-field coils. These arms

Fɪɢ. 5-2. Single-geared drum-type electric elevator machine with cam-operated contacts on the drum shaft.

are adjusted so that No. 1 is set nearest the core and *H* is the farthest away. The coil is connected across the armature. Therefore, the voltage will be increased across its terminal as the potential increases across the armature terminal, producing a stronger magnetic pull, which will draw in the contact arms one by one, depending upon the distance they are located away from the coil. This type of controller operates on the counter-electromotive force principle.

Cams operate contact arms placed at the end of the drum shaft, as shown at *A* in Fig. 5-2. Contacts at cams III and IV (Fig. 5-3) close first and control direction of motor rotation. These contactors operate diagonally; that is, contact 6 on cam III and contact 8 on cam

IV close at the same time. Assume that these two contactors control the up motion of the car; then contacts 5 and 7 will control the down motion.   Both the contactors on cam I close second; contactor 1 completes the brake-coil circuit and 2 the accelerating-magnet-coil circuit.   Both contactors on cam II close after those on I; these contactors close the main-line or potential-switch coil circuit by completing the circuit through the magnet coil and safety-device circuit, which includes the safety switch in the car, slack-rope switch, governor switch, upper- and lower-limit switches, and door contactors. If any of these switches are open, the potential switch cannot close, and the motor will not start.   When the machine is stopped, contacts on cam II open first, causing the main-line or potential switch

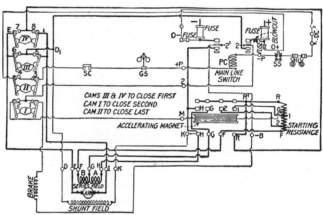

Fig. 5-3.   Diagram of the controller shown in Fig. 5-1, with all switches in off position.

to open, and the brake is then applied, bringing the machine to a full stop.   The brake shoes are raised by an electromagnet M (Fig. 5-2), and are applied by heavy spiral spring S.

Stopping at Terminal Landings.   When the car reaches the upper or lower end of the hoistway, there are three limit devices which stop it if anything happens to the operator or goes wrong with the machinery or car while it is running.

Stop balls, clamped to the hand rope at the top and bottom landings, are the first stopping device.   These balls strike a fitting, fastened to the car top, through which the hand rope passes and performs the function of the operator by pulling the operating rope up or down, depending upon the motion of the car.   Should this safety device fail, traveling nut B (Fig. 5-2) on the drum-shaft end strikes one of the fixed nuts *C* to bring the machine to rest. The third safety device is the hoistway upper- and lower-limit switches, which are in series with the magnet coil on the potential switch. Opening the circuit of this magnet coil causes the switch to drop out and open the power line. These safety stops are located at the upper and lower end of the hoistway, to prevent the car from being jammed into the overhead work or striking the bottom of the hoistway.

Controller Operation. When the operator in the car pulls the hand rope down to start in up-motion, the cams close the contactors, as in Fig. 5-4; that is, cam IV closes contact 8, cam III contact 6, cam II contacts 3 and 4, and cam I contacts 1 and 2. As previously explained, contacts 3 and 4 close last and complete the circuit from the

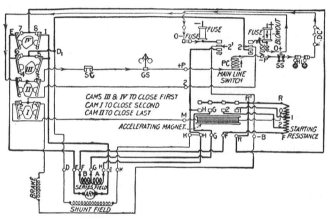

Fig. 5-4. Same as Fig. 5-3, but with potential-switch coil circuit closed.

+ side of the line, through slack-rope switch *SC*, governor switch *GS*, potential-switch coil *PC*, safety switch *SS* in the car, upper- and lower-limit switches *HL* in the hoistway, door contacts *DC*, and back to the − side of the line, as indicated by the arrowheads.

With this circuit energized the potential switch closes, as in Fig. 5-5, which completes the motor, brake-coil, and accelerating-magnet-coil circuits. The armature circuit is from the + line, through the potential-switch blowout coil, through the heavy conductor to contact 6 on cam III, from here to terminal *E* on the armature, through the armature to terminal *I* and back to contactor 8, which is closed, then to terminal *R* on the control board, through the starting resistance to terminal *F* on the series-field coils at the motor, through the series-field winding to terminal *H*, and back to the − line. The shunt field is energized from a tap taken from the wire at cam III, at point D_1, from here to terminal D on the motor, through the shunt-field winding to terminal K on the control board, which connects with terminal H, and to the $-$ line.

The brake-coil circuit is from contact 2 on the $+$ side of the potential switch to contact 1 on cam I, through the brake-magnet coil to

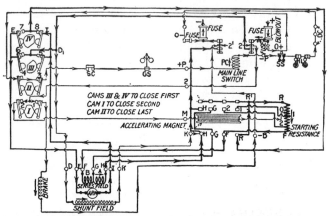

FIG. 5-5. Same as Fig. 5-3, with the controller circuits closed for one direction of elevator operation.

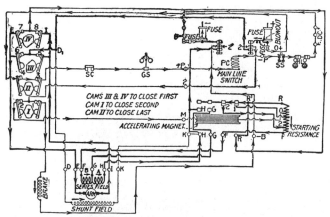

FIG. 5-6. Same as Fig. 5-3, but controller is in on position with contacts 1 and 2 closed on the accelerating magnet.

terminal $-B$ on the control board, and to contact 2' on the $-$ side of the potential switch.

The accelerating-magnet-coil circuit is completed through contact 2 on cam I, and current flow in the circuit is from contact 2 on the potential switch to contact 2 on cam I, to terminal M on the controller board, then through the accelerating-magnet coil to O on the starting resistance, through the starting resistance to terminal F on the controller, the series-field coils to H, and to — on the potential switch.

With the circuits closed, the motors should start and increase in speed; then, as the voltage across the armature terminals increases, pull of the accelerating magnet increases and resistance contact 1 closes. This short-circuits part of the starting resistance out of the armature circuit and causes the motor to increase its speed; then contact 2 closes and short-circuits another section of the starting

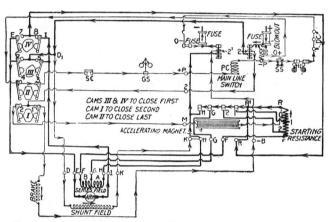

FIG. 5-7. Same as Fig. 5-3, but controller is in on position with all starting resistance and series-field windings cut out for full motor speed.

resistance. The current path through the controller with resistance-contact arms 1 and 2 closed is as shown in Fig. 5-6. In this case the circuits are practically the same as in Fig. 5-5, except that instead of the armature current flowing through all of the starting resistance, it only passes through the section between 2 and F, the circuit being from R to R' to contact arm 2 on the accelerating magnet and to point 2 on the starting resistance to terminal F of the series field, through the series field to terminal H and to the — line. The motor is now running at nearly full speed, as a compound machine with a small section of a starting resistance still in the circuit.

The next contact arm to be drawn in is G; this will cut out series-field coil B and the remaining portion of the starting resistance, which will further increase the speed of the motor; then contact arm H closes. This cuts out series-field coil A, and the motor is now running as a shunt machine at full speed.

The direction of the flow of the current is shown in Fig. 5-7 and is from the + line to contact 6 on cam III, to terminal E of the armature, through the armature to terminal I, to contactor 8 on cam IV, from here to terminal R on the controller board to R', then through contacts 1, 2, G, and H to the − line.

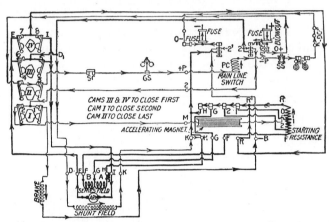

Fig. 5-8.   Same as Fig. 5-7, except that the motor-armature circuit has been reversed to change direction of motor rotation and elevator travel.

Reverse Operation.   To reverse motor rotation the cams are thrown to the opposite position, as shown in Fig. 5-8.   Contacts 5 and 7 on cams III and IV are closed instead of contacts 6 and 8.   Tracing the circuit will show them to be the same as in Fig. 5-7 except that current direction through the armature is reversed, as indicated by the arrowheads.   At the top and bottom landings the car will be stopped automatically as previously explained.

CHAPTER 6

Two-speed Direct-current Motor Controllers

Number of Speed-control Points. Elevators operating at a car speed of 100 fpm or less can be controlled with a single-speed controller without any slowdown. At higher speeds slowdowns must be provided in the control equipment so that the operator can reduce the speed of the car to make accurate landings.

For example, if a car operates at 150 fpm, the control equipment can be arranged to give two speeds, say 75 and 150 fpm. The motor on such an equipment would operate at a speed, with all the starting resistance and the series-field winding cut out, to give a car speed of 75 fpm. To obtain the 150 fpm car speed, the motor-field current is reduced to give this speed. Another method of obtaining two speeds is to use two shunt-field windings, one of which is cut out of circuit to obtain the high car speed. The higher the car speed the greater the number of speed-control points that must be provided. Some controllers have seven or more speed points on the car switch.

Circuits in the Controller. The Gurney controller (Fig. 6-1) is a full-magnetic type designed to give two car speeds from the car switch. The purpose of the different contactors is given under Fig. 6-1, and the controller functions will be made clear by a description of the circuits in Fig. 6-2, which are for a traction-elevator machine.

On this controller the up- and down-direction switches U and D also act as a line switch to make and break the line circuit, and they are equipped with blowout coils B, B_1, and B_2 to assist in interrupting the arc quickly at the contactors. Coil B_1 is between the two contactors and is effective on whichever contactor is in use.

Assume that the car switch is closed to point 2 for up motion. Closing this contact completes the circuit through the coil of the up-direction switch U, from the $+$ line to the throw-over switch A and to 1 on the car switch. This part of the circuit is common to all four circuits made through the car switch. From 1 on the car switch

119

the circuit is to 2 and to 2 on top hoistway limit switches, returning to 8 at the bottom of the controller, through the coil on the up-direction switch U, through part of resistance R_d at the bottom of the contactor D', and to 6 at the bottom of the controller. From 6 the circuit continues to 6 on the car-gate switch, thence through the safety and

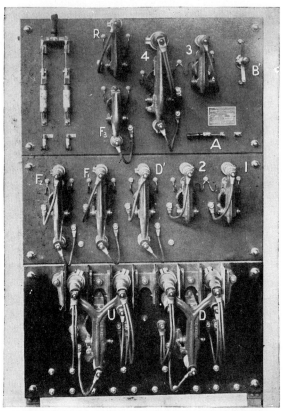

Fig. 6-1. Two-speed d-c elevator-motor controller. U, up-direction switch. D, down-direction switch. Contactors 1, 2, 3, and 4 cut out the armature resistance. Contactors F_1, F_2, and F_3 cut in the field resistance. D, dynamic-braking and noninterference magnet. R, high-speed relay. B, control switch for operating elevator at slow speed from control panel. A, switch to change from car-switch to panel-board control.

emergency switches, to 7 at the bottom of the controller, and to the — line. This circuit can be easily followed on the simplified diagram in Fig. 6-3.

Energizing the coil on the up-direction switch causes this switch to close and completes the motor circuit. The armature circuit for the motor is from the $+$ line, through blowout coils B_1 and B, the top left-hand contact of switch U, and the coil on the overload relay OL, to A_1 on the motor. The circuit is then through the armature and back to A_2 on the controller, through all the starting resistance between A_2 and R, to the top right-hand contact on direction switch U, and to the $-$ line.

The field circuit is completed from contact 12 on the up-direction switch to 12 on the motor through the field winding and back to 13 on the control board. If the inrush current through the overload relay OL is sufficient to close this relay and hold it closed, the shunt-field circuit is from 13 at the bottom of the controller through the overload relay and to the $-$ line. If the overload relay is open, the shunt-field circuit is from 13 at the bottom of the control panel, through the bottom contacts of field relays F_1, F_2, and F_3 to the $-$ line.

A circuit is provided for the brake coil from terminal 12 on the motor, through the brake coil to 10 at the bottom of the controller, through the bottom contact on contactor No. 4, to 10 on direction switch U, to the top right-hand contact of this switch, and to the $-$ line. Energizing this circuit releases the brake and allows the motor to start.

Cutting Out Starting Resistance. The starting resistance is cut out of circuit by contactors 1, 2, 3, and 4. The coils of these contactors are in series and connected across the armature, so that their closing is controlled by the counterelectromotive force developed by it. A circuit for these coils is made from the left-hand side of the up-direction switch to a and through resistance R_a and coils 1, 2, 3, and 4 to A_2 on the starting resistance, through all the starting resistance to the right-hand side of the up-direction switch and the $-$ line.

One side of the armature is connected to the left-hand side of the reversing switch and the other to A_2 on the starting resistance. These are the two points to which coils 1, 2, 3, and 4 connect. Since these coils are connected across the armature, they will have a low voltage impressed on them at the instant of starting and will not close their contactors. As the armature increases its speed, the voltage across its terminals increases, and when it reaches a certain value, the pull of coil No. 1 becomes sufficient to close its contactor.

Closing contactor No. 1 connects R_1 on the starting resistance to the right-hand side of the up-direction switch and cuts out the section of resistance between R and R_1. This increases the speed of the motor and the voltage across its terminals, and the current in coils 1 to 4 reaches a value that causes No. 2 to close its contactor. Closing

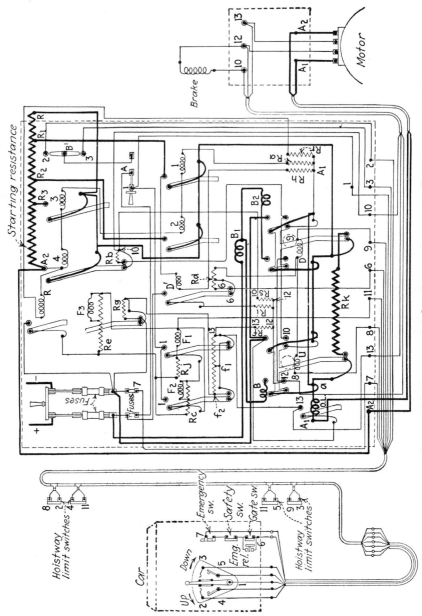

Fig. 6-2.   Wiring diagram for the two-speed d-c elevator-motor controller shown in Fig. 6-1.

this contactor cuts out the section of resistance between R_2 and R_1, which further increases the speed of the motor and the voltage across the armature and causes coil 3 to close its contactor. This latter operation short-circuits the section of starting resistance between R_2 and R_3, which causes the speed of the motor to increase further and contactor 4 to close and connect A_2 on the starting resistance to the right-hand side of the up-direction switch. Therefore, closing No. 4 contactor cuts all the starting resistance out of circuit, and the motor comes up to speed with this starting resistance cut out of circuit.

When contactor 4 closed, it opened its bottom contact and connected resistance R_b in series with the brake coil, reducing the current in this coil. Resistance R_s is connected from 12 on the left-hand side of the up-direction switch to 10 on resistance R_b. The brake coil also connects to these two points. Therefore, resistance R_s is in parallel with the brake coil. When it is disconnected from the line, its inductance sets up a current through resistance R_s. This current through the brake coil, which exists for a very short period, tends to retard its application and prevents mechanical shocks when stopping the car. Connecting a resistance in parallel with the brake coil also prevents a high voltage from being induced in it when it is disconnected from the line.

High-speed Operation. A circuit for relay R is provided from the left-hand side of the direction switch through resistance R_r and coil R to A_2 on the starting resistance. Closing contactor 4 completes a circuit from A_2 to the — line for coil R, and this relay closes to establish a circuit for field contactors F_1, F_2, and F_3, when the car switch is closed to No. 4 or 5 contacts. Closing the car switch to the high-speed point 4, when relay R is closed, completes the circuit for the coil of field contactor F_1, from 1 on the car switch to 4 and then to 4 and 11 on the top hoistway-limit switches, back to 11 on the control panel, through coil F_1 and the contact on relay R, to the — line. When contactor F_1 closes its top contact, it opens its bottom contact. The latter connects the resistance between 13 and f_1 in series with the field coils and causes an increase in speed.

Closing the top contact on switch F_1 completes the circuit for F_2 coil, from throw-over switch A to the top contact of F_1 through coil F_2 and resistance R_c, and to the — line. This causes F_2 contactor to close and open its bottom contact, which introduces the section of resistance between f_1 and f_2 into the field circuit, causing a further increase in speed. When the top contact of F_2 closed, it completed the circuit for F_3 coil and caused this contactor to function and open its bottom contact. Opening bottom contact of F_3 connected resistance R_g in series with the field coils, and the motor now comes up to full speed. The field resistance is cut in by three steps, which gives a smoother acceleration than if it were introduced in one step.

Controller Operation during Stopping. The dynamic-brake contactor D' has its coil connected from the left-hand side of the up-direction switch, through resistance R_h, to the top contact of contactor 1. When contactor 1 closes, its top contact is connected to the right-hand side of the up-direction switch and to the — line. Therefore, when contactor 1 closes, in addition to cutting out a section of the starting resistance, it also connects coil D' directly to the line and

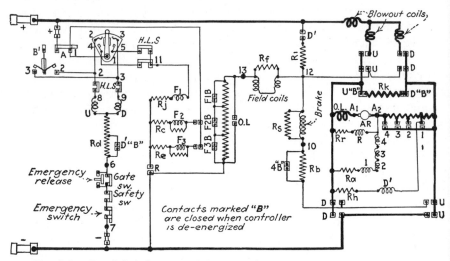

Fig. 6-3. Simplified diagram of the control connections shown in Fig. 6-2.

it closes its top contact. This provides a circuit for the field coils from the + line to the top contact of relay D' through resistance R_i and through the field coils back to 13 on the controller and to the — line. Consequently, as long as D' remains closed the field coils are connected to the line through resistance R_i, whether or not the up-direction switch is open. The purpose of this circuit is to retain a magnetizing current in the field coils to cause the armature to produce a dynamic-brake action to assist in stopping the elevator.

Coils F_2 and F_3 are connected in parallel with part of resistances R_c and R_e, respectively, while coil F_1 is connected in series with resistance R_j only. If all the coils were connected alike, the contactors would all open at about the same time and the resistance would be cut out of the field coils in practically one step, causing an abrupt slowdown. The way the coils are connected to their resistance, contactor F_1 opens almost instantaneously when the car switch is moved off the high-speed point. When coil circuit F_2 is interrupted by opening the top contact of F_1, the current in the coil will not immediately decrease to zero. Owing to inductance a current will be maintained through the coil and the resistance connected to its terminals. This will cause a short delay in opening contactor F_2. A similar delay is caused in opening F_3, so that a gradual slowing down of the motor is obtained when stopping.

When the car switch is centered, the direction switch opens; then through the dynamic-brake action of the motor and the application of the mechanical brake the elevator is stopped. Opening of the up-direction switch connects resistance R_k between the bottom contacts of the two direction switches and completes a circuit for the armature, from A_1 on the motor to the bottom contact of the up-direction switch, through resistance R_k to A_2 on the starting resistance, and back to A_2 on the motor. This circuit is based on the assumption that contactors 1, 2, 3, and 4 have remained closed. As previously explained, the coils of contactors 1, 2, 3, and 4 are connected across the armature. Therefore, they hold their contactors closed until the armature has come down to a slow speed.

Contactor D' holds closed until the motor reaches a slow speed and, in addition to maintaining a current in the field coils for dynamic-braking action, it also acts as a noninterference magnet. As long as the bottom contact of D' is open, all of resistance R_d is in series with the coils of the direction switches. Even if the car switch is moved to the on position, sufficient current cannot flow through the direction-switch coil to close its contacts as long as all of R_d is in circuit. Contactor D' is so designed that it will hold closed until the motor has slowed down below where contactors 1 to 4 open. This arrangement prevents the operator in the car from suddenly reversing the motor before the starting resistance has been cut back into circuit.

When the overload relay OL is used, if a sustained overload is applied to the elevator, this relay will close and, as previously shown, short-circuit the field resistance. Therefore the motor can be operated at slow speeds only. Closing this relay on the first inrush of current to the motor ensures the motor having full field at starting.

Resistance R_f is connected from 12 on the up-direction switch to 13 on the field resistance. These two points are the field-coil terminals; consequently this resistance is connected in parallel with these coils. When the car is inched to the floor or similar movements are being made, relay D' may not have time to close before the direction switch is opened, in which case the field circuit is broken. It is to prevent high voltages from being induced in the field coil under these conditions that resistance R_f is connected across the field coils. When the field circuit is broken, the induced voltage causes a current to flow through resistance R_f and prevents the voltage from exceeding normal limits.

Stopping at Terminal Landings. When the car is to stop at the terminal landings, the hoistway-limit switches function to stop the car in a way similar to the way centering the car switch does. First a cam on the car comes in contact with the first hoistway-limit switch and opens it. Opening this switch performs the same function in slowing the car down as moving the car switch off point 4. Further movement of the car opens the last hoistway-limit switch, which has the same effect as centering the car switch, and the car comes to a stop.

In the foregoing the operation of the controller has been considered for only one direction of the car. The reverse motion is the same as for up except that down-direction switch D closes instead of up-direction switch U.

CHAPTER 7

Direct-current Traction-machine Rheostatic Controllers

Traction-elevator Machines. These machines have practically superseded drum types. Controllers for their motors, particularly for high car speeds, offer special problems, and many types have been developed. These may be divided into two general classes: rheostatic and variable-voltage. Before the latter came into general use for high-speed elevators, smoothly accelerating and slowing down the machine with rheostatic control led to many interesting developments. One of these is the six-speed Otis control shown in Fig. 7-1.

Six-speed Controller. With the car switch shown in Fig. 7-2 and the controller shown in Fig. 7-1 six car speeds are obtainable, which are: direction, 1, 2, 3, 4, and 5, respectively. At the terminal landings cams are installed in the hoistway which engage a roller secured to an arm on the stopping switch (Fig. 7-3) mounted on the hitch beam, on top of the car. Operation of the stopping switch opens a series of contacts which reduce the speed in four steps and then stops the elevator by releasing the direction switch. The automatic stopping switch on top of the car is set so that its contacts drop out in the reverse order in which they close. There are seven contacts for each direction on the automatic stopping switch, as shown in the upper left-hand corner of Fig. 7-4, and the last three contacts, Nos. 5, 6, and 7, open simultaneously. Numbers 5 and 6 each open one side of the direction magnet, and No. 7 opens the circuit which parallels the brake magnet, providing for a quick release of the brake at the terminals.

In addition to the circuits controlled by the automatic stopping switch, these are the final limit switches for overtravel at the terminals, the switch in the car for emergency, safety-plank switch under the car, which opens when the safeties are set by action of the

127

governor, and a governor switch in case of overspeed. Any of these will release the potential switch, which opens the power line to the machine and brake circuit and applies a hard dynamic brake by means of connecting resistance across the armature.

Controller Circuits. The wiring diagram in Fig. 7-4 is of the controller shown in Fig. 7-1, looking at the panel from the rear. The

Fig. 7-1.   Front of Otis MFL4C controller for gearless traction machines.

contacts are shown in a position assumed with the main switch open. As each switch operates, the contacts shown open are closed and those that are closed immediately open. With all emergency switches closed the elevator may be operated on any speed by moving the car switch to any desired position. The car switch is generally wired so that moving its handle away from the side of the hoistway that has the most openings will cause the elevator to run up, and moving it the other way will cause the elevator to run down.

When the main-line switch is closed, as in Fig. 7-4, the shunt-field circuit is immediately energized, which provides a positive field on the motor at the instant of starting. This circuit is from $+$ line to D terminal at the bottom of the controller, through the shunt-field winding on the motor to the K terminal on the controller, through the SFR resistance to the $-$ terminal on the potential switch L, and to the $-$ terminal at the bottom of the controller.

Potential-switch Coil Circuit. After the main-line switch is closed, if all emergency devices such as the governor switch, final limits, safety-plank switch, emergency switch in the car, etc., are closed, the potential switch will close. All emergency switches with the exception of the last-named are normally closed, since these open only when called upon by overspeeding of the car, overtravel at the terminal landings, etc. The emergency switch is operated at will by the operator and is generally opened when the operator leaves his car.

Top main contacts M_a of the potential switch close the main-line circuit to the controller. These contacts are equipped with blowout coils to prevent arcing. Auxiliary contacts A_c complete the line to the various operating and brake circuits. Bottom main contacts M_b connect the section of bypass resistance between E' and I across the armature to insure a quick stop should the potential switch be released when the elevator is in motion. Lower auxiliary contacts A_l are used to short-circuit the resistance in series with closing coil L, which ensures a strong pull when closing the switch from the off position. After the switch pulls in, these contacts open to cut resistance in series with the magnet coil and reduce the current.

An overload magnet O_m connects in series with one power line to actuate a contact in series with the potential-switch closing coil should the current exceed a predetermined amount.

A small single-pole double-throw knife switch S_p is on the left of the potential switch. With the single-pole switch in the down position, the potential switch will not close automatically when the line switch is closed, but will remain closed after it has been closed by hand provided all emergency switches are closed. With the single-pole switch in the up position, as in Fig. 7-4, the potential switch closes automatically when power is on the board, provided all emergency switches are closed.

General practice provides for the potential switch to be self-closing. The circuit for energizing closing coil L on this switch is from its top main contact marked $+$ through the switch S_p and the small fuse to L terminal at the top of the board, then into the conduit C_w and through the L and $+L$ contacts on the car-governor switch. From the governor the circuit returns to the top of the board and goes into the C_v conduit and up through the counter-weight-governor switch, when used, and one set of contacts on the limit switch at the top of the hoistway, then to $+L''$ on the bottom limit switch through the slack-rope switch, when used, on the compensating sheave, back to the $+P$ terminal on the controller, down to resistance *ARP,* through its short-circuiting contacts A_l, coil L,

Fig. 7-2.   Six-speed car switch with cover removed.

and the contacts on overload relay O_m, to $-P$ at the top of the board. From the top of the board the circuit again goes to the top hoistway-limit switch and through contacts $-P$ and $-L'$, then through contacts $-L'$ and $-L$ on the bottom hoistway-limit switch to the $-L$ terminal on the safety switch in the car.   From the Y terminal on the safety switch in the car the circuit goes to the switch mounted on the safety plank under the car and from here to the Y' terminal on the board, then through a fuse and to the $G' -'$ contacts on the car-governor switch, returning to the $-'$ terminal on the board and the $-$ top contact of the potential switch, thus completing the circuit.   This causes the potential switch to close and complete the power circuit to the controller. A simplified diagram of the potential-switch circuit is shown in Fig. 7-5.

The operation of the control will be considered under the following conditions: throwing the car switch completely to full-speed position up and stopping automatically at the top terminal landing. Any other conditions of operation will merely change the time rate of acceleration and stopping, with the exception of the brake operation, which will be explained for the condition of lifting full load to the top terminal landing.

Direction-switch Coil Circuits. With the car-switch handle moved to full-speed up position, the circuit is completed for the up-direction switch B on the board, and it closes. This circuit is from M_a contact on the $+$ side of the potential switch through the heavy line to tap X on the left-hand side of the board. From tap X the circuit

Fig. 7-3. Stopping switch that is mounted on top of the car.

is through a fuse, $AR4$ resistance to terminal SL' and through the hoistway-door contacts, back to SL terminal on the board, then to between the No. 5 contacts on the stopping switch mounted on top of the car. From $+U$ on the stopping switch the circuit returns to $+U$ on the controller, through closing coil B on the up-direction switch, N_r contacts on the nonreversing magnet G, to U' terminal at the top of the board. The circuit now goes to U' on No. 6 contacts of the stopping switch and to U and C on the car switch, then to C and $O-$ on the safety switch back to $O-$ on the board, and down through a fuse and the right-hand auxiliary contacts A_c on the potential switch to the $-$ line, thus causing the up-direction switch to close.

The pulling in of the up-direction switch closes the contacts shown open and opens those shown closed in Fig. 7-4. The up- and down-direction switches are mechanically interlocked so that the action of one direction switch when pulling in causes the other switch to drop back to the extreme position open, closing the bottom con-

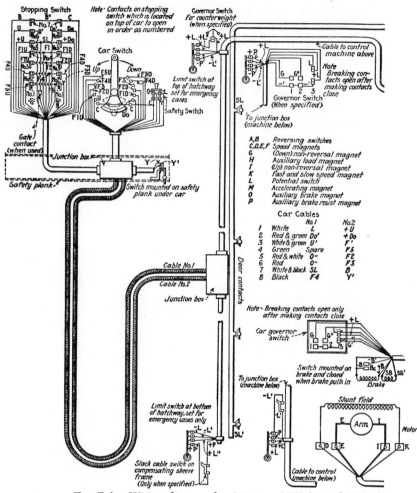

FIG. 7-4.   Wiring diagram for Otis type MFL4C and MFH4C mag-

tacts. These switches are so interlocked to prevent the bottom contacts on both switches from closing at the same time, for should this
happen, the armature would be short-circuited through them, causing a very severe stop.   Closing the up-direction switch completes
the circuit to the motor's armature, from the M_a terminal on the
+ side of the potential switch through 2 and 4, and 1 and 3 contacts in parallel on the up-direction switch, to the left and down
through the right-hand bottom contact of down-direction switch,
which is closed when the up-direction switch closes.   Then the circuit is through hold-down coil of down-direction switch to *I* through

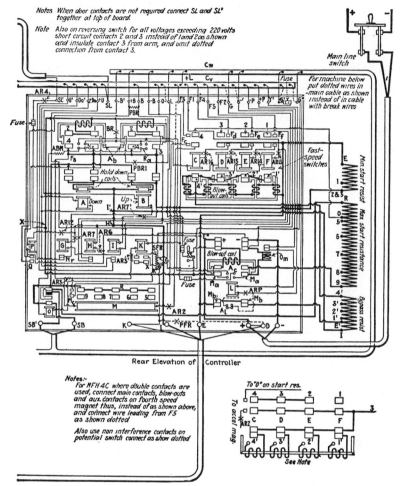

netic controller for medium-speed gearless traction elevator machines.

the armature and back to *E* on the board, through the minimum
starting resistance and line marked 2 and 3, back through the left-
hand bottom contact on down switch to *SB* terminal, through the
series winding of the brake back to *SB'*, through the closed contact
of auxiliary brake magnet *O*, to *R* on the maximum starting resistance, to 9, through right-hand M_a contact of the potential switch,
and through the overload magnet to the − line.

Releasing the Brake. The auxiliary making contacts, R_a, on the up-
direction switch *B*, when closed, complete the circuit to the shunt
winding of the brake from the + line through the main contact on

potential switch L, through the left-hand auxiliary contact A_c, through the up-direction switch auxiliary contacts R_a, the BR resistance, to the $+B$ terminal, then through the shunt coil of the brake magnet back through the $-B$ line and small fuse, through the other auxiliary contact A_c on the potential switch, to the $-$ line. The brake magnet then releases the brake shoes from the pulley and allows the motor to start on slow speed.

The initial lift of the brake is obtained by the combined action of the shunt and series windings. The action of the brake lifting closes contact B_c mounted upon it, which provides a circuit through the auxiliary brake magnet O on the controller, from the bottom contacts R_a on the up-direction switch, through ABM resistance, coil O, to $-B'$ terminal, and to the B_c contacts on the brake back to the $-B$ terminal on the board to the $-$ line. Auxiliary brake magnet coil

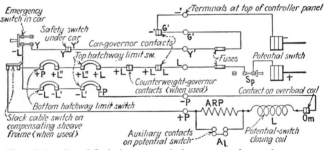

Fig. 7-5. Simplified diagram of the potential-switch circuit.

O closes, which first short-circuits the series winding of the brake and then open-circuits the line leading to this winding. This eliminates any sneak current through the series winding due to poor short-circuiting contact. A sneak current through the series winding of the brake would change the character of its action and should be avoided.

It has been shown that the armature is in series with the minimum and maximum starting resistance and the series winding of the brake at starting, the series winding of the brake being short-circuited by O magnet after the brake has lifted. In addition to this circuit there is a bypass circuit across the armature to obtain a positive slow speed which could not be obtained otherwise, as load conditions would greatly vary the low speed with the starting resistance in series with the armature.

Resistance Paralleled with Armature for Low Speed. This bypass circuit is from a tap I' just below and between the direction magnet coils, to the *I* terminal on the bypass resistance, through this resistance to 4′, through the bottom contact of the fast-speed magnet coil 4′ (*C*), through the blowout coils of fast-speed magnet coils *C, D, E,* and *F,* to *E* lead of the armature. From here the circuit is completed to the − line as explained for the armature.

The fast- and slow-speed magnet *K* operates immediately when either direction switch pulls in. It is connected in series with resistance, which is mentioned here particularly because the resistance is utilized for other means than for reducing current value (see Fig. 7-6). The contact which *K* operates short-circuits the shunt-field resistance of the motor. The final point on the car switch completes a short circuit across this magnet, causing it to release, cutting in the shunt-field resistance and getting final high speed of the motor. This cannot be done, however, until after the accelerating magnet *M* contacts pull in and the auxiliary contact M_c on the last arm of

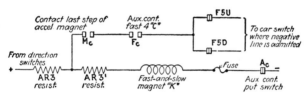

Fig. 7-6.   Simplified diagram of shunt-field resistance magnet circuit.

the accelerating switch has closed. The circuit for the fast- and slow-speed magnet *K* is from a tap F_s just below contact 4 on the down-direction switch, down through *AR3* and *AR3′* resistances, coil *K,* to the − on the potential switch. Closing magnet *K* contacts short-circuits the *SFR* resistance in the shunt-field circuit, providing full field on the motor during the starting period. The conditions so far described are those which would obtain if the car switch closed contacts *U* and *C,* to give the slowest operating speed. With the car switch in full on position a number of other operations take place to bring the motor to full speed. The second, third, fourth, and fifth operations in accelerating the motor are closing fast-speed magnet switches *C, D, E,* and *F.*

Operation of Fast-speed Switches. The first fast-speed switch to close is *F,* and the circuit for its coil is from the auxiliary contact R_a on the up-direction switch, through coil *F* and *AR13* resistance, to terminal *F1* at the top of the board, to between No. 4 contacts on the stopping switch, to *F1U* contact and to *F1U* on the car switch, and to the − line, as previously explained.

Fast-speed switch F, closing, cuts in the portion of the bypass resistance between 1′ and 2′ by opening its lower contact. The top main contact, closing, short-circuits the portion of minimum starting resistance between E and 1. The auxiliary contact F_f completes the circuit for fast-speed magnet E.

It should be noted that lower contacts 1′, 2′, and 3′ on the fast-speed magnets connect to 1′, 2′, and 3′, respectively, on the bypass resistance; also, top contacts 1, 2, and 3 connect to terminals 1, 2, and 3, respectively, on the minimum starting resistance, although the connections are not shown.

A circuit to close the second fast-speed magnet switch E is obtained from $+$ line as for F, through coil E, resistance $AR14$, auxiliary contact F_f on F, to the $F2$ terminal on the board. From here the circuit is to between No. 3 contacts on the stopping switch to contact $F2U$ on the car switch and it returns to the $-$ line as previously explained. Energizing coil E causes contacts 2 to close and 2′ to open. Closing contacts 2 short-circuits the remainder of the minimum starting resistance, and opening contacts 2′ cuts in the portion of the bypass resistance between 2′ and 3′, thus increasing the motor speed. Closing auxiliary contacts F_e completes the circuit for fast-speed magnet coil D, which is the fourth step in accelerating the motor. A circuit is obtained for coil D from the same line as E, through D, resistance $AR15$, auxiliary contacts F_e, to $F3$ terminal on the board, and then directly to car-switch $F3$ contact, and to the $-$ line.

The circuit for D magnet coil does not pass through the stopping switch, but goes directly to the car switch. As this circuit is used when the motor is under load at the terminal landings, it will be further considered when the stopping at the terminal is discussed. The top main contact of D switch short-circuits the minimum starting resistance in the same manner as the E switch; in fact, the contacts are paralleled. The bottom main contact inserts bypass resistance between 3′ and 4′, and auxiliary contact F_d completes the circuit to fast-speed magnet C.

Closing magnet coil C is the fifth step in starting the motor. A circuit for this magnet coil is from the $+$ line as for D, through coil C, resistance $AR16$, to terminal $F4$, to between the No. 2 contacts on the stopping switch, then to $F4U$ on the car switch, returning to the $-$ line. Closing switch C opens contacts 4′, which open-circuits the bypass resistance. Closing contacts 4 closes the circuit to the accelerating magnet M at the bottom of the board and closing auxiliary contacts F_c provides a circuit to be completed for the final speed only after the accelerating magnet M has closed all its contacts. Operation of the accelerating magnet, which then follows, is entirely automatic upon the closing of C magnet contact.

Accelerating-magnet-coil Circuit. One side of the accelerating magnet is connected to the top main contacts of the direction switches at F_s, where it obtains the $+$ line, and the circuit continues through the accelerating magnet coil M through $AR2$ resistance, then through top contact 4 of fast-speed magnet switch C to point O in the starting resistance, the latter being connected to the $-$ line of the main circuit and to the armature, so that the circuit is completed from O to 9 and to the $-$ line. This connection provides for any degree of acceleration desired between the minimum and maximum time required for the accelerating magnet to close its five contacts 5, 6, 7, 8, and 9. By the shifting of O closer to R the accelerating magnet connection is brought closer to the armature; that is, the voltage across coil M is more dependent upon the speed at which the armature rotates. The speed of the armature is approximately a measure of the voltage drop across its terminals, and when the accelerating magnet is connected across the armature terminals, the strength of magnet M will increase as the speed of the motor. Shifting connection O toward 9 increases the voltage drop across M, making it more independent of the speed of the armature until, when O is connected at point 9, full line voltage is obtained across the accelerating magnet circuit, at which point magnet M pulls in its contacts rapidly and entirely independently of the motor acceleration.

Accelerating magnet M, upon operating, closes its five contacts in sequence, each short-circuiting a portion of the maximum starting resistance. Contacts 5, 6, 7, 8, and 9 on M connect two points, 5, 6, 7, 8, and 9, respectively, on the maximum starting resistance. Contact No. 5 is the first to close and No. 9 is the last.

Connecting Resistance in Shunt-field Circuit. The last contact arm of the accelerating switch closes an auxiliary contact M_c in addition to the one which short-circuits the last portion of starting resistance, and this auxiliary contact provides for releasing the fast- and slow-speed magnet K. It was shown that K magnet coil obtained the $+$ line from contacts on the direction switch, through $AR3$ and $AR3'$ resistances, through K coil, small fuse, auxiliary contact on the potential switch, and to the $-$ line. To release this magnet, the $-$ line passes from C to $F5U$ contact finger in the car switch, to $F5U$ lead on No. 1 contact in the stopping switch, to $F5$ on the controller, through auxiliary contact F_c on fast-speed magnet C, auxiliary contact M_c on the last arm of M, to a point between $AR3$ and $AR3'$ resistances in the *K* magnet-coil circuit. Under this condition, the — line connects to both sides of the *K* coil and is equivalent to short-circuiting this coil, causing it to release its contactor (see Fig. 7-6). The contacts on *K* magnet open the short circuit across *SFR* (shunt-field resistance), as shown in Fig. 7-4. This weakens the field, giving the last and final speed of the motor.

The elevator is then running at full speed, and should there be a tendency to overspeed because of light load up or full load down, the governor closes contact *G*, which short-circuits a portion of the *SFR* resistance, between — and *G*, to maintain correct speed. If this does not hold the speed within a predetermined limit, contacts *G″* on the governor switch close and short-circuit another section of the *SFR* resistance between *G* and *G″*.

Stopping at Terminal Landings. When a terminal landing is approached, contacts in the stopping switch (Fig. 7-4) open in the following order:

No. 1 opens and causes *K* switch to close by removing the short circuit from across its coil, and again short-circuits the *SFR* field resistance, increasing the field strength and slowing down the motor.

No. 2 opens and causes fast-speed switch *C* to release and open its top main contact, thus opening the circuit to the accelerating magnet which, releasing, cuts the maximum starting resistance in series with the armature. Closing the bottom main contacts of *C* connects the bypass resistance across the armature.

No. 3 opens and causes fast-speed switch *E* to release, which closes its bottom contacts and short-circuits a portion of the bypass resistance 2′ and 4′. The top main contacts open, but do not remove the short circuit on a portion of the minimum starting resistance, since these contacts are in parallel with the top main contacts on switch *D*. Opening of auxiliary contacts on switch *E* releases switch *D*. When fast-speed switch *D* releases, opening its top main contacts removes the short circuit from the section of minimum starting resistance between 1, 2, and 3. Closing the bottom contacts of *D* does not have any effect since the section of bypass resistance between 2′ and 4′ has already been short-circuited by switch *E* closing contacts 2′. Therefore, the only effective operation has been to open the short circuit across minimum-starting-resistance terminals.

No. 4 on the stopping switch opens the circuit through fast-speed switch *F*, which releases and opens its top contacts, removing the short circuit across the section of minimum starting resistance from *E* to 1. The bottom main contact short-circuited the portion of bypass resistance between 1′ and *E′*. The elevator will now run at its minimum speed with all starting resistance in series with the armature and the minimum amount of bypass resistance across the armature terminals.

Should the load be heavy, the motor might be unable to raise it to the top landing with all resistance in series and the minimum amount of bypass across the armature. Should this be the case, the auxiliary load magnet H, which is connected across the armature leads from H' on the I lead through coil H and the $AR6$ resistance to E'' on the E lead, is so adjusted that if the armature slows down greatly, H magnet releases. The reason for this was outlined when it was explained that the voltage across the armature terminals was approximately a measure of its speed. The releasing of H magnet closes its contacts, which permits F magnet to close again and short-circuit a portion of the minimum starting resistance; and cuts in that part of the bypass resistance between $1'$ and $2'$, increasing the voltage drop across the armature and giving a greater torque for the motor to raise the full load to the top-floor landing.

The circuit for fast-speed magnet coil F is from $+$ on the potential switch through coil F and $AR13$ resistance as before, but instead of the circuit going to F_1, it now goes through the contacts on H, back to $F3$ terminals at the top of the board, to $F3$ contact on the car switch, to C contact, and returns to $-$ line.

Numbers 5, 6, and 7 contacts on the stopping switch open simultaneously. Each of Nos. 5 and 6 opens one side of direction-switch magnet coil. Number 7 contact opens the parallel or discharge resistance PBR across the brake magnet coil. This provides for a quick positive stop at the terminals. The releasing of the direction switch disconnects the armature and brake from the line, at the same time releasing K magnet, which opens its contacts and inserts SFR resistance in the shunt-field circuit.

Preventing Overtravel at Terminal Landings. Should the car overtravel the terminal landings, it will open either the top or bottom shaft-limit switches, which release the potential switch. When the top contacts of this switch open, power is cut off from the controller; then when the auxiliary contacts A_c open, the brake and operating circuits also open. The closing of the bottom contacts of the potential switch connects the section of the bypass resistance between E' and I across the armature, and causes a strong dynamic-braking action to stop the elevator as long as the armature is in motion.

Down nonreversing magnet coil on switch G is so connected that only when down-direction switch A is closed is coil G connected across the armature. This contact opens the up-direction-switch circuit and prevents reverse travel until the armature has practically come to rest.

Up nonreversing magnet coil I is connected so that only when up-direction switch B is closed is coil I connected across the armature. The function of this magnet is similar to G, but is used for the opposite direction of travel.

There are two parallel circuits across the shunt winding of the brake. One of these circuits PBR is of high resistance; the other has low-resistance $PBR1$. At the terminal landings these parallel circuits are disconnected by the stopping switch on the top of the car. This circuit is opened by contact No. 7 in the stopping switch, one contact provided for each direction of travel. Figure 7-7 is a simplified diagram of the brake circuits. High-resistance circuit PBR can be disconnected only during regular operation by the stopping switch.

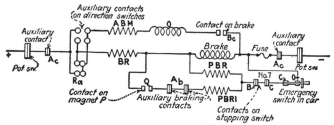

FIG. 7-7. Simplified diagram of the brake-coil circuits.

Low-resistance circuit $PBR1$ is taken through the auxiliary contacts A_b on the direction switches and contact Q on the auxiliary brake-resistance magnet P. Since the auxiliary brake-resistance magnet is connected across the armature, contact Q is closed only when the car is traveling at high speed; then when the car-switch handle is centered, auxiliary contacts A_b on the direction switches will close before the elevator comes to rest. This permits the brake magnet to discharge through both parallel circuits when stopping from high speed.

When the car stops from low speed, such as when inching or making a short run, the auxiliary brake-resistance magnet will not close its contact. Therefore, the brake magnet can discharge through the high-resistance PBR only, thus insuring quick brake application.

Governor Prevents Overspeed. The governor has two making and two braking contacts and in this case will have four settings, all of which will be overspeed travel. The first setting closes a contact G which short-circuits a portion of shunt-field resistance between —

and *G*. This should check the speed of travel, but if it is not effective, the second making contact *G″* short-circuits an additional portion of shunt-field resistance, giving a further check to the speed of travel. Such conditions as overload, falling elevator, open shunt-field winding, or poor connections at definite places would prevent the governor making contacts *G* and *G″* from producing the desired results, in which case the third setting, contacts *G′* and *L*, provides for the opening of the potential-switch magnet coil. When the potential switch releases, it provides a hard dynamic stop by the elevator motor, disconnects the power from the equipment, and opens the brake circuit. The fourth setting of the governor provides for applying the car safeties. This action is rarely obtained, as the only condition which should call for it is when the elevator car is mechanically disconnected from the elevator machine, which is only when hoisting ropes have broken.

It must be remembered that the controller described in the foregoing is for a definite purpose and that this description applies to this control only. The reader by a little study can understand any control of a similar type, as the general principles are similar.

CHAPTER 8

Rectifiers Used with Alternating-current
Motor Controllers

Electromagnets. Beginning on page 56, Chap. 1, the use of a-c motors on elevators and their controllers is discussed. Here several control methods will be considered. At best, most a-c electromagnets are not highly satisfactory. They are generally noisy in operation and take a high current inrush to start closing; if they fail to close, their coils overheat quickly; and they are in general larger and more costly than comparable d-c designs. To eliminate these objections most new a-c elevator-motor controllers are now powered by direct current supplied through metallic-disk rectifiers of selenium or other type. Some control manufacturers now build only d-c controls and brakes for a-c elevator motors, having abandoned their a-c designs.

Rectified Brake-coil Current. Figure 8-1 diagrams a Haughton push-button control for a small geared drum-type elevator driven by a three-phase across-the-line-start squirrel-cage motor. This equipment has a d-c brake coil powered by a metallic-disk rectifier.

Motor and brake-coil circuits are at the top of the diagram with the brake-coil rectifier-transformer connected across two of the motor terminals. The rectifier is represented as four triangles connected in series. Each triangle represents a group of rectifier disks. These rectifiers have a low resistance to current flow from the triangle base to its apex and a very high resistance in reverse direction. To energize the coil and release the brake, contacts *BT*, *B2*, and *B3* are closed.

When motor contacts close, the rectifier-transformer is energized to supply current to the brake coil. A circuit is from the top transformer-secondary terminal through contact *B3*, thence down through top center section of the rectifier, contacts *B2* and *BT* the brake coil, to the bottom section of the rectifier and to the bottom transformer

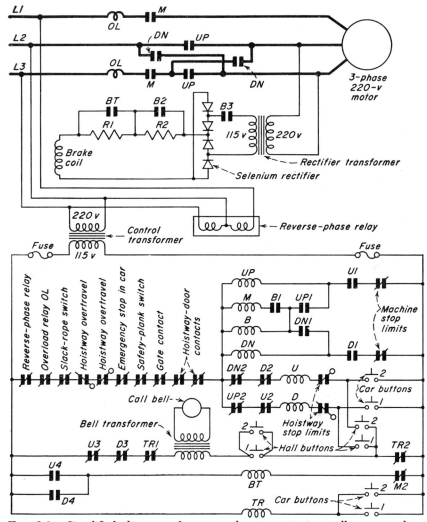

FIG. 8-1. Simplified diagram of an a-c elevator-motor controller operated on alternating current with a d-c brake supplied through metallic-disk rectifiers.

terminal. When transformer polarity reverses, the circuit is up through the bottom center section of the rectifier, contacts B2 and BT and the brake coil as before, to the top section of the rectifier and through contact B3 to the transformer's top terminal. Even though transformer polarity reverses, current flow through the brake coil is in the same direction.

Contact BT is closed by a timing relay and opens after the brake has released. Opening contact BT puts resistance R1 in series with the brake coil, reducing the current in it and the load on the rectifier. Contacts $B2$ and $B3$ open simultaneously, $B3$ to disconnect the brake coil, permitting the brake to be applied by a spring. Opening $B2$ puts 3,000-ohm resistor $R2$ in circuit to protect the rectifier against high induced voltage from the brake coil by holding reverse current to a low value.

Reverse-phase Relay. A reverse-phase relay protects the elevator against motor reversal. It has happened that after the power supply lines were disconnected, a phase has been reversed when the lines were reconnected. This reverses all polyphase motors inside the reversed connection, a very dangerous condition with geared drum-type elevators. The car or counterweights may be pulled into the overhead work and the equipment wrecked.

Push buttons are of the two-pole type. Poles 2 and 2 of the car buttons are parts of the same unit and are closed simultaneously when the second-floor button is pressed in the car. Likewise, Nos. 1 and 1 poles of the car button are also two poles of the same button. This is also true of the hall buttons. For simplicity, one pole of No. 1 button has been shown in the diagram (Fig. 8-1) with one pole of No. 2, but they are parts of two separate units.

Control Operation. The control is for two-floor operation with constant-pressure push buttons. Assume that the car is at the bottom floor and that No. 2 car button is pressed and held closed. Bottom No. 2 pole closes a circuit through timing-relay coil TR to open contacts $TR1$ and $TR2$ in the call-bell and hall-button circuits to isolate them during up-direction operation.

Closing the top pole of No. 2 button completes a circuit through up-relay coil U. If we assume that the left-hand terminal of the control transformer is $+$, a circuit for U will go from the transformer through the several limit and safety switches, contacts $DN2$ and $D2$, coil U, normal-stop top limit switch, and button 2 to the transformer. Energizing relay coil U opens contact $U2$ in the down-direction relay coil D circuit and $U3$ in the hall-button circuit, and closes $U1$ in the up-direction-switch coil circuit and $U4$ in the brake-timing relay coil BT circuit. Energizing coil BT closes contact BT in the brake-coil circuit to short-circuit resistor $R1$.

Closing contact $U1$ energizes up-direction switch coil UP, from the transformer, through the limit and safety switches, coil UP, contact $U1$, and machine stop limit, and back to the transformer. Coil UP opens contact $UP2$ in down-direction relay coil D circuit and closes contact $UP1$ in main-switch coil M circuit and the two UP contacts in the motor circuit. However, the motor cannot start until the M

contacts close. Closing contact *UP1* completes the brake relay coil *B* circuit. This relay closes contacts *B2* and *B3* in the brake-coil circuit to release the brake, and closes *B1* in the main-switch coil *M* circuit. This switch closes the two *M* contacts in the motor circuit to start the elevator. Switch *M* also opens *M2* contact in the brake-timing relay coil *BT* circuit to deenergize this coil. Opening contact *BT* in the brake-coil circuit is delayed for a short time by the timing action of the relay, to allow the brake to release before putting resistor *R1* in the brake-coil circuit.

As the car approaches the top floor, a cam on it opens the up hoistway-limit switch in the up relay coil *U* circuit to stop the car at the floor. If for any reason the car does not stop, the machine stop limit in the up-direction-switch *UP* circuit opens; if the car continues to overtravel, a cam on it opens the top hoistway overtravel switch to break all control circuits.

When the up button in the car is released it opens its two poles. The bottom pole breaks the circuit through timing relay coil *TR*, but contacts *TR1* and *TR2* are held open a short time by the timing action of the relay. This allows time to open the car gate and hoistway door to prevent the car from being started by the hall button at the lower landing.

Down Direction. After the car has been cleared at the top floor and the car gate and hoistway door close, the car can be started from the lower landing by pressing the hall button, which in this case is No. 1. Right-hand No. 1 hall button closes a circuit for down-direction relay coil *D* through button No. 1 and timing relay contact *TR2*. No. 1 hall button on the left closes the call-bell transformer circuit through contacts *U3*, *D3*, *TR1*, and *TR2*. If the hoistway-door, car-gate, and other safety contacts are closed, direction relay coil *D* is energized, closing contact *D1* in the down-direction-switch coil *DN* circuit and opening contact *D2* in the up-direction relay coil *U* circuit. Down-direction relay also opens contact *D3* in the call-bell transformer circuit to prevent this bell from ringing. If the car-gate and hoistway-door contacts are not closed, direction relay *D* cannot open contact *D3* and the call bell rings, calling attention to the fact that the second-floor hoistway door and car gate are not being closed properly and the elevator cannot start down.

After the down relay *D* closes, the control functions in about the same way as for up direction, except that the *DN* contacts instead of the *UP* contacts close in the motor circuit to reverse the current in one phase of the motor and reverse its direction of rotation and car travel. To simplify the diagram (Fig. 8-1) the control has been shown for two-floor operation, but by adding hall buttons and hoisting-door contacts it is possible to adopt the control to any number of floors. However, at the intermediate floors the car must be stopped by releasing the push button and the hoisting doors must be of a design through which the car or floors can be seen.

D-C Controllers. Figure 8-2 is a simplified diagram for a Westinghouse three-phase a-c squirrel-cage elevator motor with two steps of resistance in the stator circuits. Control which is from a car switch is on direct current supplied by a copper oxide rectifier. Motor wiring with its starting resistance shown at the top of Fig. 8-2 has the motor protected by thermal overload relays TO. Below the motor comes a reverse-phase relay RP and then the three-phase rectifier, connected to the control and brake circuits through two fuses.

Timing Relay. Just below the rectifier are coils N and $N1$, which are parts of timing relays M and $M1$. Coils N and $N1$ are in series, connected directly across the line, and are energized at all times when power is on the rectifier. Timing-relay coil M is also connected across the line through contact $P2$, which is closed. This circuit is from $+$ on the rectifier, to coil M and its parallel resistance $R2$, to contact $P2$, to $-$ on the rectifier. This energizes timing-relay coils M and N, which open contact M in accelerating-contactor coil $A1$ circuit. Timing-relay coil $M1$ circuit is opened through contact $P3$ and this relay cannot close even though its N coil is energized.

Car-switch Operation. Contacts U and D for this switch are in the lower right-hand corner of Fig. 8-2. Assume that the car is at the bottom landing and the car switch is moved to the up position, closing contact U. This completes a starting circuit from $+$ on the rectifier, through terminal overload contact TO, reverse-phase relay contact RP, the governor switch, terminal overtravel limits OTU and OTD, the stop button in the car, car-gate and hoistway-door contacts, accelerating-switch contacts $A1$ and $A2$, power-contactor coil P, up-direction contactor coil UP, contact DN, up-limit switch UL, car-switch contact U, to $-$ on the rectifier.

Energizing coils P and UP close contacts P and UP in the motor circuits and the motor starts with all resistance in series with its windings. Let us assume that lines $L1$ and $L3$ have $+$ instantaneous polarity and $L2$, $-$. Then current flows in through the top and bottom motor windings and resistors and out the center resistor and winding to $L2$. Closing contactors P and UP also closes contacts $UP1$ and $P1$ in the brake-coil circuit to release the brake. The motor now starts smoothly with all resistance in circuit.

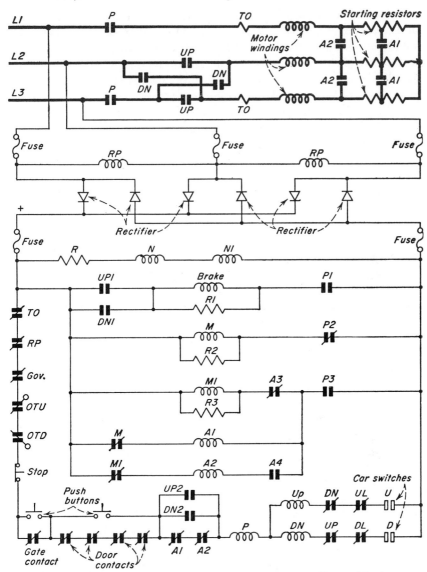

FIG. 8-2. Simplified diagram of an a-c elevator-motor controller and brake operated on direct current supplied by metallic-disk rectifiers.

When contactor *P* closed it also opened contact *P2* in timing-relay coil *M* circuit and closed *P3* in timing-relay coil *M1* circuit. Relay *M1* opens contact *M1* in accelerating-contactor coil *A2* circuit so that this contactor cannot close when contactor *A1* closes contact *A4* in *A2* coil circuit.

Accelerating Circuits. Opening contact *P2* in timing-relay *M* coil circuit causes it to start timing out and in a few seconds closes contact *M* in coil *A1* circuit. Making this coil alive closes contacts *A1* to cut out part of the resistors in the motor circuit and increase its speed. When contactor *A1* closed it opened contact *A3* in timing-relay coil *M1* circuit and closed contact *A4* in accelerating-contactor coil *A2* circuit. Opening contact *A3* in relay coil *M1* circuit starts this relay timing out, which closes contact *M1* in *A2* coil circuit. Contact *M1* closes in a few seconds to energize coil *A2*, which closes contacts *A2* to short-circuit the resistors and bring the motor and elevator smoothly to full speed.

As the elevator car approaches the top floor a cam on it opens top limit switch *UL*. This breaks the circuit through contactor coils *P* and *UP* to cut power out of the motor and brake-coil circuit, and stops the car level with the top floor. If for any reason the car runs by the top floor it opens overtravel contact *OTU*, opening the circuit through all motor contactor coils to stop the motor. Once the car runs onto an overtravel limit it cannot be started again from the car switch, but must be moved manually either by holding in the correct motor switches or by turning the brake wheel with a spanner wrench.

Accelerating contactors *A1* and *A2* each have a closed contact *A1* and *A2* in series with motor contactor coils *P, UP,* and *DN*. Direction contactors *UP* and *DN* each have a normally open contact *UP2* and *DN2* in parallel with *A1* and *A2* contacts. This arrangement of contacts ensures that either one of the motor direction contacts *UP* or *DN* must be closed before the accelerating switches can close and remain closed. When accelerating switch *A1* closes, it opens contact *A1*, which breaks the circuit through the motor contactor coils unless either contacts *UP2* or *DN2* have been closed by one of the direction switches closing.

A constant-pressure push button connects in parallel with the gate contact and the hoistway-door contacts. If the gate or hoistway-door contacts fail to close, holding in the corresponding button permits moving the car to where the trouble can be easily corrected. If the elevator fails to start closing these buttons will also show if the trouble is in the gate or door contacts or somewhere else in the equipment.

When the car switch is moved to down position it closes contact *D* to complete a circuit through power contactor coil *P* and down-direction contactor coil *DN*. Closing contacts *DN* in the motor circuit reverses current direction in its two bottom windings, to reverse its direction of rotation. From here the control functions in practically the same way as for up direction.

D-C Operation from Rectifiers. Small elevators in buildings where alternating current only is available may be driven by d-c motors powered from rectifiers. Figure 8-3 is a simplified diagram of such an equipment built by Keystone Electric and operated from a car switch. Rectifiers and a three-phase power supply are shown at the bottom of Fig. 8-3, with the motor armature at the center and the car switch and control relays at the top.

When the rectifier is energized from the a-c line, current flows from + terminal of the rectifier through the potential-switch coil *P* circuit to — on the rectifier. This circuit includes the top final limit *TFL* in the hoistway, bottom final limit *BFL*, governor switch *G*, stop switch in the car, motor overload contact *MO*, and the rectifier thermal overload contact *TO*. The potential switch closes contacts *P* and *P* in the d-c power circuits to the elevator motor and its control.

As soon as contacts *P* close, the motor shunt-field coils are energized through resistor *R8* and the field protective relay coil *FPR*. This relay closes contact *FPR* in the control circuits at the top of the diagram. Unless the shunt-field circuit is complete, contact *FPR* will not close, so that the motor armature cannot be energized unless the shunt field is alive.

Down-direction Relay. Assume that the car is at the top landing when the car switch is closed to down position *D*. This completes a circuit through the down relay coil *SD* and the hoistway-door and the car-gate contacts. Relay *SD* closes contact *SD*, which completes a circuit through the brake relay coil *BR* and down-direction contactor coil *D*. Brake relay *BR* closes contacts *BR* in the brake-coil circuit to release the brake. Contact *BR1* closes across resistor *R8* to put full voltage on the motor shunt-field coils and the field relay *FPR*. Down-direction contactor *D* closes contacts *D1* and *D2* in the armature circuit. This circuit is from + on the rectifier, through blowout coil *BO* on the potential-switch contact *P*, blowout coil *BO* on contact *D1*, motor overload coil *MO*, the armature, contact *D2*, starting resistors *R5* to *R7*, motor series-field winding, to — on the rectifier.

Accelerating-contactor coils *A1*, *A2*, *A3*, and *A4* are connected across the armature through contact *SD4*, closed by down-direction relay coil *SD*. As the armature accelerates, contactor *A1*, followed by *A2*, *A3*, and *A4*, closes as the voltage across the armature increases. These contactors cut out starting resistors *R5*, *R6*, and *R7*

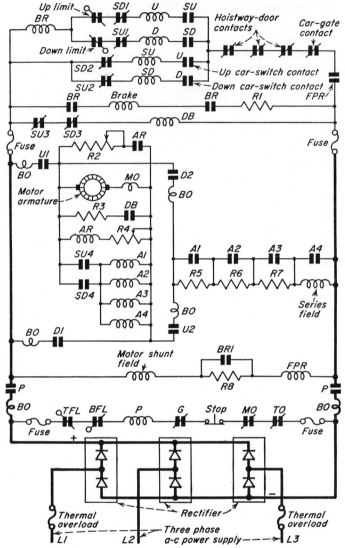

Fɪɢ. 8-3. Simplified diagram of a d-c elevator motor and controller powered from an a-c system through metallic-disk rectifiers.

in succession to bring the motor to full speed; then contact *A4* closes to short-circuit the series-field winding to convert the motor to a shunt machine, as explained in Chap. 5, page 112. The elevator is now operating at full speed.

As the car approaches the bottom floor a cam on it opens the down limit switch in series with down-direction contactor coil *D* and the brake relay coil *BR*. This opens contacts *D*1 and *D*2 in the armature circuit to cut power out of the motor, and to open contacts *BR* in the brake-coil circuit to stop the car at the floor.

Dynamic Braking. When down-direction contactor closed it opened contact *SD*3 in the dynamic-braking contactor coil *DB* circuit to prevent contact *DB* from closing in the dynamic-braking resistor *R*3 circuit. During stopping, when armature contacts *D*1 and *D*2 opened, contact *SD*3 closed to energize coil *DB* and close contactor *DB*, connecting dynamic-braking resistor *R*3 across the armature. The motor now becomes a generator and supplies current from its right-hand terminal through overload coil *MO*, contact *DB*, and resistor *R*3 to the left-hand armature terminal. Current flowing in this circuit causes the motor to act as a brake and assist the mechanical brake in stopping the car at the floor.

Regenerative Braking. When the elevator is lowering a heavy load it has a tendency to overspeed. If powered from a d-c outside source or through a motor generator the elevator motor can become a generator and supply current into the power system. This re-generative action causes the motor to produce a braking effect that keeps the elevator and its load under control. Current can flow freely in only one direction through metallic-disk and electronic recti-fiers. Consequently, if the elevator tends to overspeed, its motor cannot regenerate back into the power system to keep its speed under control.

In an attempt to solve this problem, speed-control resistor *R*2 is provided, as shown in Fig. 8-3. Contactor coil *AR* is connected through adjustable resistor *R*4 across the armature. If the elevator starts to overspeed, voltage generated by the motor becomes high enough to close speed-control contactor *AR*. Current now flows from the right-hand terminal of the armature through overload coil *MO*, contact *AR*, and part of resistor *R*2 to the left-hand armature terminal. Regenerative current flowing through the armature pro-duces a braking effect to help hold the elevator speed near normal. This method is not all that could be desired, because its successful operation depends on the closing of a relay. If the relay fails to function the elevator may overspeed.

Up Direction. To operate the elevator in up direction, contact *U* is closed by the car switch to energize up relay coil *SU*. This relay opens contacts *SU*1 in down contactor *D* coil and *SU*2 in the down relay coil *SD* circuits to isolate them and closes contact *SU* in up contactor and brake relay *BR* circuit. Relay *BR* closes contacts *BR*

CHAPTER 9

Unit Multivoltage Signal Control with Microleveling[1]

Unit Multivoltage Control. Several types of automatic signal-control system, with automatic leveling at landings for elevators, have been developed by the Otis Elevator Company, but type 11SLU, described in this chapter, is typical. Control of elevator motor speed is by regulating generator voltage; unit multivoltage control (variable voltage) and leveling are done by the main elevator motor. Power for the elevator is supplied by a motor generator comprising a three-phase 220-volt squirrel-cage motor, a d-c compound generator, and a compound-wound exciter. The generator has two shunt-field windings, one for normal operation and the other for leveling the car into the floors. The motor windings are arranged for star connection during starting and idling periods, and for delta grouping when the car is in motion; the control is shown in Fig. 9-1.

Control Panel. With this control (Fig. 9-2), the car is stopped and leveled to floors automatically, either by the car attendant pressing buttons in the car corresponding to floors called by passengers or in response to buttons pressed at landings by waiting passengers. To start the car, the attendant moves the starting switch D to start position (Fig. 9-3). When the hoistway and car doors close, the car starts and accelerates to full speed. After the car starts, the car switch may be centered, and the car will continue in motion until it approaches a floor for which a button has been pressed. It will then slow down and level into the floor, when the car and hoistway doors will open automatically.

Floor Selector. Floor selectors are required on all automatic-type elevators. On the more modern Otis signal-control elevators the selector performs three important functions: it stops the car at floors

---

[1] Harry F. Cater and William Devaughn assisted in writing this chapter.

153

for which buttons have been pressed, levels the car at floors, and operates signals necessary for efficient elevator service.

The selector (Fig. 9-4), may be considered a miniature elevator traveling in a hoistway in which the operating equipment is conveniently grouped. A traveling crosshead *C*, representing the elevator, is driven vertically by a steel tape attached to the car and

Fɪɢ. 9-1. Alternating-current three-phase motor-generator starter for the elevator controller shown in Figs. 9-6 and 9-10.

wound on sheaves at the top of the hoistway. A connection between the tape-sheave shaft and vertical screw, driving the selector crosshead, is made by a chain drive and reduction gears. Pressure or any hall or car button energizes a corresponding stationary contact on the selector. Brushes on the crosshead pick up the signals as the car approaches the desired floors, and stopping operation is then initiated on the controller. The final stop at floor level is controlled by selector cams and contacts.

On the selector are so-called floor boards *B*, (Figs. 9-15 and 9-17), one for each floor at which stops are to be made. These floor boards are straight bars on which are mounted the up and the down con-- tacts for initiating car stops at floors; contacts for illuminating the hall lanterns, to operate lights in the hall and the car-position indi- cator; and other signal units. The floor boards are clamped to three vertical bars so that they may be adjusted vertically.

At the top of the selector shown in Fig. 9-4, which is for a type 11SLU control, are two sets of contacts (Fig. 9-16). On the right the two rows, with four contacts each, are for various circuits in the control system and the slowing down of the elevator to microleveling speed. Contact $1U$ controls the microleveling-device magnet; $1D$ is in the hall-light relay circuit; $2U$ and $2D$ are in the first high-speed- switch coil circuit; $3U$ and $3D$ are in the generator-field and brake- switch-coil and the auxiliary generator-field and brake-switch-coil circuits; and $4U$ and $4D$ are in the car- and hoistway-door operator circuits.

These contacts are closed by weights and opened by movement of the selector crosshead. Two vertical rods R and R_1 (Fig. 9-4), con- nect to the contacts at their top ends and are connected at their lower ends by a rod and angles (Fig. 9-13), so that when one rod moves up, the other goes down. On these rods are U-bolts U, one on each rod for each floor at which stops are to be made. As the car approaches a floor at which a stop is to be made, the pawl magnet P (Figs. 9-4 and 9-14) is deenergized and the pawls are allowed to drop out and engage a U-bolt corresponding to the floor at which the stop is to be made. As the car slows down, the selector crosshead continues to move and, through the pawl engaged with the U-bolt, moves the vertical rods to open the two rows of contactors shown at the right of Fig. 9-16. Thus the car is gradually brought to micro- leveling speed.

The four contacts on the left of Fig. 9-16 are for microleveling, two for high-speed and two for low-speed leveling. The two high- speed leveling contacts H are closed by magnet Mc when the car starts. Contact MU is for up-direction and MD down-direction leveling. These contacts also connect to two vertical square rods V that pass through two roller arms A carried on the selector cross- head (Figs. 9-4 and 9-15). When coil Mc is energized, the square rods are rotated to separate the roller arms to clear the leveling cams L. These cams are mounted on a shaft driven from the same gearing as the crosshead screw for control leveling at the floors.

During stopping, when $1U$ contact opens, it breaks the solenoid coil Mc circuit, and the leveling contacts are released. If the level-

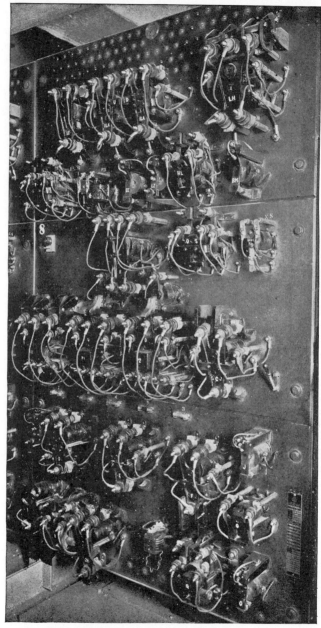

Fig. 9-2.　Otis type 11SLU panel for microleveling signal control.

FIG. 9-3.   FIG. 9-4.

FIG. 9-3.   Car operator's panel for microleveling signal control. *A*, nonstop switch. *B*, slow-speed switch. *C*, emergency-stop switch. *D*, operator's start switch. *E*, microcontrol switch. *F*, emergency switch to short-circuit door contacts. *G*, car-light switch. *H*, reverse switch. *I*, red light to show when motor generator is running. *J*, key switch for starting motor generator.

FIG. 9-4.   Floor selector, parts of which are identified in the text.

ing contacts were free, the mechanism that controls them and is attached to solenoid *Mc* would open the high-speed contacts, close the direction contacts, and then open the latter.   However, when solenoid *Mc* releases the contacts, assuming that the car is in up direction, the roller arms are allowed to close, except for the one for up-direction leveling, which is prevented from doing so by a leveling cam.

This cam stops the microcontrol mechanism in a position in which the up high-speed contact remains closed and the up-direction contact closes. The down-direction leveling contacts open. As the leveling cams continues to rotate it first opens the up high-speed contact and then the up-direction contact as the car stops level with the floor, if the equipment is properly adjusted. When the car stops, the roller arms on the direction leveling contacts just clear their cam, so that if the car moves only a small distance above or below the floor, the slow-speed leveling contact will be closed to bring it back level with the landing.

In this description of the control equipment a motor-generator set, the motor of which operates on three-phase power, will be considered. At the top of Fig. 9-6 is shown a straight-line diagram of a type D starting equipment for such a motor. As previously mentioned the motor windings are arranged for star connection during starting and idling periods and for delta grouping when the elevator is operating. Change from one connection to the other is made automatically as the elevator is started and stopped.

Starting the Motor Generator. The switch in the car for motor generator starting and stopping is a lock type. To start the motor generator, a key is inserted in the lock and turned clockwise, which closes contacts A and B, contacts C and D remaining closed. Closing contacts A and B completes a circuit for the line-switch relay coil LL, if the single-pole knife switch on the control panel and the three overload relay contacts, $O1$, $O2$, and $O3$, are closed, as shown in Fig. 9-6. Overload relays $O1$ and $O2$ are in the motor circuit, and $O3$ is in the generator and elevator-motor connections, all three being of the lockout type.

The starting circuit is from L_2 line, through fuse No. 5, contacts $O1$, $O2$, and $O3$, knife switch S, idling-switch relay coil LL, contact Ma, and to line L_3. Energizing coil LL closes contact LL in line-switch coil L circuit. Making this coil alive pulls in the two-pole line switch L, opens contact La in the elevator-motor shunt-field circuit, and closes contact La in the potential-switch coil C circuit. The motor is now connected star to the line and it starts and comes up to speed. PM contact in the start-switch holding circuit is closed by coil PM, connected across motor winding T_3-T_6. The motor circuits at one instant are from the L_1 line, through switch L and motor winding T_1-T_4, to the connection between M_1 switches. Another circuit is from the L_3 line, through motor winding T_3-T_6 and left-hand switch M_1, and to the circuit from line L_1, joining this circuit. The two circuits then go through the right-hand M_1 switch and motor winding T_2-T_5 to the L_2 line. It will be seen that the star connection for the motor windings is formed between switches M_1 and M_1.

As the motor comes up to speed, the exciter builds up its voltage and energizes coil KR, contact LL having been closed when the line-switch relay LL closed. Making coil KR alive closes contacts KR in the start-switch and light-in-car circuits. Closing KR_1 contact completes a circuit for the lamp in the car, which lights to show that the motor has started and that exciter voltage has built up. KR_2 contact makes the starting-switch holding circuit through contacts CD so that the starting key may be removed and the motor generator continue running.

Elevator-motor Shunt-field Circuit. A circuit for the elevator-motor shunt field is from D_1, through the field winding with resistance 1FPR in parallel, resistance 1$FR1$, contact F_2, shunting resistance 1$FR2$, and relay coil J (auxiliary contact La having been opened when motor switch L closed), to — on the exciter. This excites the elevator motor with a strong field at starting and closes relay contact J in the potential-switch circuit. Relay J prevents potential switch C from closing until the elevator-motor shunt field is excited. If the motor field strength is reduced below a safe value, relay J will open and cause potential switch C to open, killing the control circuits and stopping the elevator.

The potential-switch circuit is from + on the exciter, through fuse 4, stop switch in the car, safety-plank switch under the car, broken-tape and compensating-rope-sheave switches, top-limit switch L_8 and bottom-limit switch L_6 in hoistway, La, governor and J contacts, potential-switch coil C, auxiliary contact Ca on potential switch, CR_1 resistance, top-limit hoistway switch L_1, bottom-limit hoistway switch L_3, fuse 3, and single-pole knife switch on control panel, to — on the exciter. Potential switch C now closes and makes the elevator-motor and generator control panel alive. Auxiliary contact Ca opens and connects CR_2 resistance in series with the potential-switch coil.

Direction-switch Control. Assume that the car is at the bottom landing ready to start in the up direction. The direction switches are of the mechanical latched-in type and are positioned by a selector contact when the car stops at the terminal landings. As the car approaches the bottom landing, contact B (Fig. 9-7) on the selector closes and completes a circuit through up-direction switch coil P_1, down-direction switch latch coil P_2, and contacts 1Ha and LHa, to

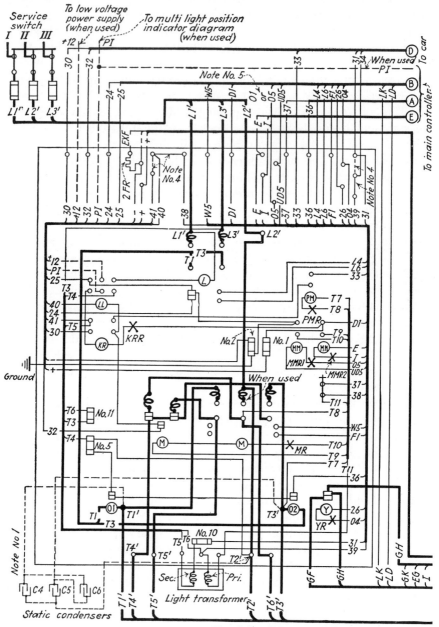

FIG. 9-5. Diagram of the controller shown

—01 line. P_1 coil pulls in up-direction switch P_1 and it latches closed. P_2 coil unlatches down-direction switch P_2 and it opens.

When P_1 switch closes, it opens contact $P1a$ and kills the coils, but up-direction switch remains closed ready for the car to start in up direction. Closing of the up-direction switch closes contacts P_1 and P_1 in the elevator-generator and motor circuit, shown at the bottom of Fig. 9-6, but these machines remain dead through contacts $2H$ and $1H$, which are open.

Initial Starting of the Car. Initial starting of the car in either direction is done by moving the car switch to run position, where it closes contacts in three circuits. One of these is for closing the car and landing doors, a second energizes in closing coil on the stop switch, and the third contact is in the generator-field and brake-

NOTES:—

1 — *When static condensers are used connect as shown dotted*

2 — *Before connecting mains to controller, measure the potential between each line and ground and then connect the line of highest potential to line Ll.'*

3 — *Markings T1 to T11 inclusive are for convenience of wireman only and are not to be marked on starter.*

4 — *For 250 volts or under connect as shown and omit dotted connections between studs 31 and 34 and between 40 and 41. for over 250 volts connect as shown dotted and omit solid connections between studs 34 and 39 and between 38 and 40.*

Marking "05" to be used for all types of main controllers except SLU and ALU controllers in which case it becomes O1

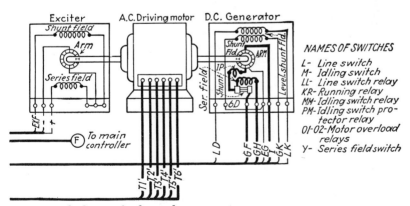

NAMES OF SWITCHES

L- Line switch
M- Idling switch
Ll- Line switch relay
KR- Running relay
MM- Idling switch relay
PM- Idling switch protector relay
O1-O2- Motor overload relays
Y- Series field switch

in Fig. 9-1, looking at back panel.

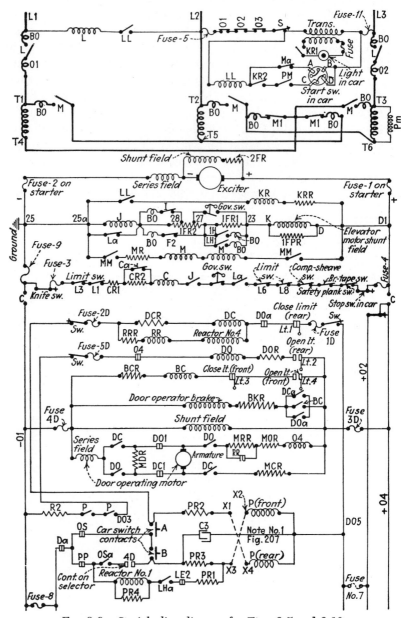

Fig. 9-6. Straight-line diagram for Figs. 9-5 and 9-10.

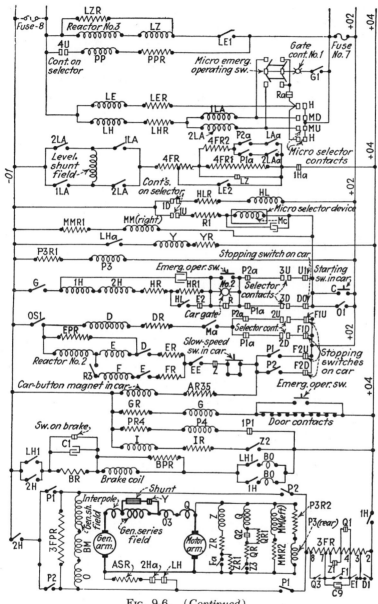

Fig. 9-6. (*Continued*)

switch coil circuit. Circuits for starting are completed by the car and hoistway doors closing.

Door-operator Control. The door-closing circuit is made by contact A on the car switch. This circuit is from $+02$ line, through fuse $1D$, limit switch Lt_1, door-open switch DOa contact, door-close direction-switch coil DC, DCR resistance, fuse $2D$, car-switch contact A, contacts OS and Da, and fuse 8, to -01 line. Contact DCa closes in the brake-coil circuit, releasing the brake to free the doors so that they may close. Contacts DC and DC close in the door-operator-motor circuit, which starts and closes the car and hoistway doors. The door-operator-motor shunt field is energized all the time by being connected directly to $+04$ and -01 lines. The armature circuit is from $+04$ line, through MCR resistance, DC contact, door-operator-motor armature, DO_1 and DC contacts, and series-field winding, to -01 line.

When the doors start to close, open limit switches Lt_2 and Lt_4 close. The latter completes a circuit for the brake and cutoff relay coil BC, which when energized closes contact BC in the brake circuit. This contact closing allows power to be cut off the motor by opening limit switch Lt_1, with the brake released by Lt_3 remaining closed, so that the door will drift closed before the brake is applied, at which point Lt_3 opens. About all that the brake does is to hold the doors in the closed or the open position.

After the doors are accelerated, a large part of their retarding is done by dynamic-braking the motor. During closing operation, when direction switch DC opens, it closes contact DC_1 and connects resistance MDR across the motor-armature terminals. The motor then acts as a brake to retard the doors and bring them nearly to rest in closed position when the brake shoes are applied.

On the car switch there are three positions, two in the run direction and one in the door-open position. Moving the car switch to the first run position would close contact A only. This would close the doors, but the car would not start. If the car switch is centered, the doors remain closed. Then, to open the doors, the car switch must be moved to door-open position. This closes contact B and makes a circuit for the front coil of P relay through resistance PR_3 and contacts B, OS, and Da, to the -01 line. Relay contacts P and P close and complete a circuit for open direction-switch coil DO.

The opening operation is practically the same as the closing. When the doors are closing, the rear coil of relay P is energized through car-switch contact A and opens P relay contacts. Several types of door operator and controls may be used with this elevator

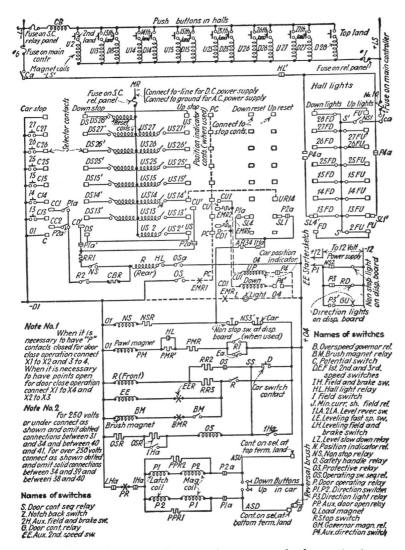

Fig. 9-7. Straight-line diagram of stopping and selector circuits.

control besides the one shown. Frequently the operator opens the doors only, and puts springs in tension to close them when the brake is released.

How the Control Functions during Starting. When contact A closed to energize the door operator, contact D (Fig. 9-7) closed a circuit for stop-switch R front coil, from $+04$ line, through OS contact, RR_2 resistance, and front R coil, to -01 line. This switch

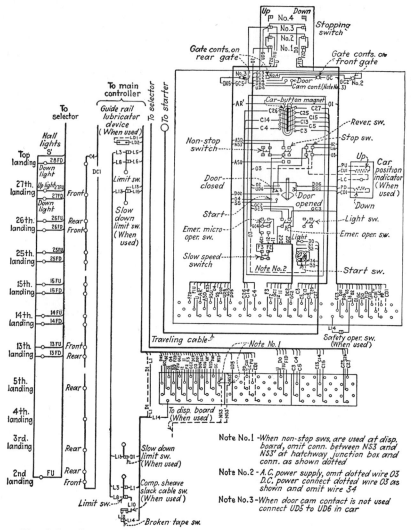

FIG. 9-8. Car-switch and push-button panel and limit-switch circuits.

closes several contacts, one of which is *R,* to complete a holding circuit for its front coil, locking the switch closed so the car switch may be centered and the control circuits remain alive. Closing car-switch contact *D* also made a circuit for the operating-switch sequence relay *OS,* but this relay cannot close until contact *1Ha* on the field and brake switch closes. When stop switch *R* functioned, it also closed contact R_1 and completed the circuit for the pawl magnet, which when energized releases contacts $1U$ to $4U$ and $1D$ to $4D$ on top of the selector, closing them (Fig. 9-16). All circuits for the control are included in Fig. 9-6 (on two pages) and in Fig. 9-7. In the following description of the circuits it will be necessary to refer quite frequently from one diagram to another.

When the hoistway doors close they complete a circuit for the door-contact relay coil G (Fig. 9-6) from $+04$ line, through door contacts, door-contact relay coil G, and resistance GR, to -01 line. The relay closes contact G in field- and brake-switch coil $1H$ circuit, and auxiliary field- and brake-switch coil $2H$ circuit. This circuit is from $+02$ line, contact C on the car switch, through contact $U1$ on the stopping switch on top of the car, selector-switch contact $3U$ (closed when the pawl magnet is energized), down-direction switch contact $P2a$, car-gate contact No. 2, contacts E_2 and HL (contact HL was closed when hall-light relay coil HL was energized by contact $1D$ closing on the selector), resistance HR, coils $2H$ and $1H$, and contact G, to the -01 line. Switches $1H$ and $2H$ close the brake-coil and the generator-field circuits. The brake is released and excitation applied to the generator.

The brake circuit is from $+04$ line, through contact $1H$, the brake coil, contact on brake, and $2H$ contact, to the -01 line. The brake is released and its contact opens, putting resistance BR in series with the brake coil. The generator shunt-field circuit is from $+04$ line, through contact $1H$, the field resistance (except that part short-circuited by contacts $Q1$ and $Z1$), contact $P1$, safety-handle relay coil O, selector brush-magnet relay coil BM, the generator shunt-field winding, and contacts $P1$ and $2H$, to the -01 line.

The generator field is made alive with low excitation and the elevator motor starts at slow speed. When relay $2H$ functioned, it opened contact $2Ha$ to isolate the motor and generator armatures from the excitation circuit. Contact Y is closed to connect the shunt across the generator series-field winding. The generator has a weak field and the motor a strong shunt field, resistance $1FR2$ being short-circuited by contact F_2 and $1FR1$ by contact $1H$; the latter closed when $1H$ relay was energized. These conditions cause the motor to operate at slow speed.

When relay coil O in series with the generator shunt field is made alive it closes contact $O1$, in parallel with the car-starting switch C. This contact maintains a circuit through relay coils $1H$ and $2H$ after the starting switch is centered.

When the brush magnet coil BM is energized it closes contact BM in the selector brush-magnet-coil circuit (Fig. 9-7). This coil ad-

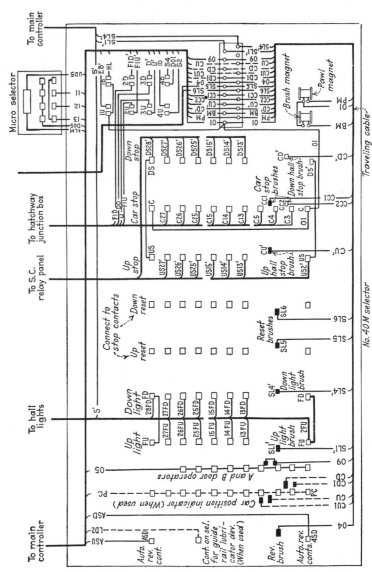

FIG. 9-9. Diagram of floor-selector connections.

vances the selector brushes in the direction that the crosshead is traveling.

When the car-starting switch was closed, it also completed a circuit through the right-hand coil MM of the idling-switch relay, on the motor-generator start panel, through resistance $MMR1$, to -01 line, causing this relay to close its two contacts. This completes the circuit for the two coils on idling switch M, which are connected across the exciter from $+$ to $-$ lines. The M switch throws over and changes the motor connections from star to delta, by opening the two $M1$ and closing the three M contacts.

The left-hand coil of the idling-switch relay MM is connected across the elevator-motor terminals. This coil holds the relay closed after the right-hand coil has been deenergized during stopping, until the elevator almost comes to rest. When switch M functions it closes auxiliary contact Ma in the first-speed relay coil D circuit. This contact prevents the elevator control from switching to high speed until the driving motor on the generator has been connected delta.

Coil EE of auxiliary second-speed switch is energized through the same R contact as the pawl magnet, and closes its contact EE in the second- and third-speed circuits E and F, respectively. When relay $1H$ closed, it made contact $1Ha$ short-circuit resistance OSR_1 and energized coil OS of operating sequence relay till it was strong enough to close its contacts, one of which, $OS1$, is in the first-, second-, and third-speed circuits D, E, and F (Fig. 9-6).

The first-speed circuit is through coil D, from $+02$ line, through $F1U$ contact on the stopping switch on the car, selector contact $2U$; $P2a$ contact on the direction switch, Ma contact on motor-generator switch M, resistance DR, coil D, and relay contact $OS1$, to the -01 line. Relay D closes two contacts. One of them, $D1$, closes across a section of generator shunt-field resistance and raises the generator voltage, causing the elevator motor to increase in speed.

The other contact on D relay completes the circuit for the second-speed relay coil E. This coil is in series with an iron-core reactor coil, to time the closing of E. When E closes it pulls in contact E_1 to cut out a section of resistance in the generator-field circuit. It also closes contact E to complete the circuit for third-speed relay coil F and opens contact E_2 across HR_1 resistance, putting it in series with coils $1H$ and $2H$.

Closing of relay F is timed by the same reactor as that of E. When relay F closes, it closes contact F_1 across a section of generator shunt-field resistance, now partly short-circuited by contacts

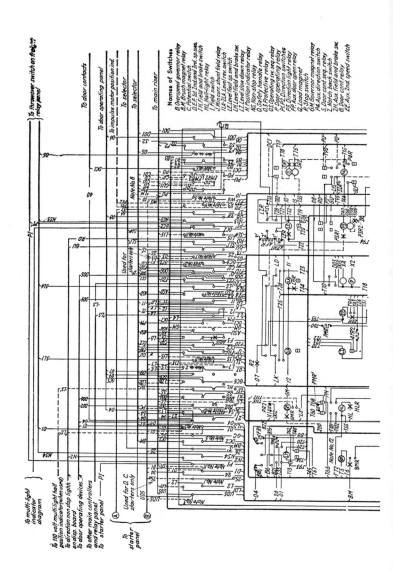

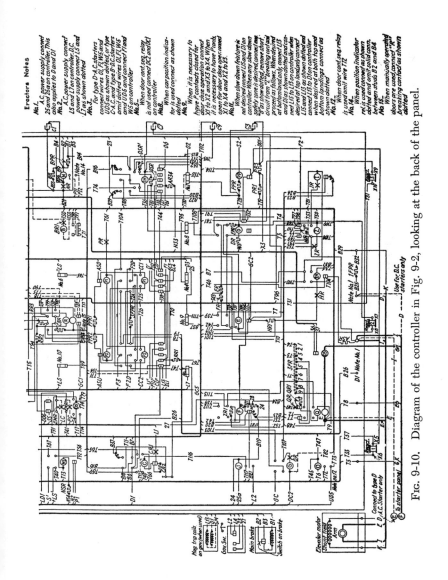

FIG. 9-10. Diagram of the controller in Fig. 9-2, looking at the back of the panel.

Q_1 and Z_1, and opens contact F_2 in parallel with the motor shunt-field resistance $1FR_2$. The motor again increases in speed, due mostly to connecting resistance $1FR2$ in series with the motor field.

When relay F pulled in, it closed contact Fa across resistance ZR_1 in series with relay coil Z. This relay functions and opens a contact across a section of generator-field resistance now short-circuited by contact F_1. Relay Z also closes a contact, Z_2, in series with field-switch coil I, and another, Z_3, in series with relay coil Q across the motor armature. Field switch I functions and opens a contact across a section of motor-field resistance $1FR2$.

Closing Z_3 contact in series with Q relay coil causes this relay to function and open a contact Q_2 in series with its coil, and connects resistance QR_1 into circuit. Contact Q_1 opens across a section of resistance in the generator shunt field already short-circuited by contact F_1. Contact Q_3 closes and short-circuits the last section of resistance out of the generator shunt field, and the motor comes up to full speed.

When the car switch was closed it also completed a circuit through coil Mc on the microselector device that controls the position of the leveling contacts while the car is in normal operation. Coil Mc closes the high-speed microselector contacts H and energizes coil LE of the fast-speed microleveling switch. This switch closes contact LE_2 across the resistance in series with the generators leveling shunt-field winding. This field-winding circuit is still open through direction contacts $1LA$ and $2LA$ and auxiliary contact $1Ha$, the latter opening when $1H$ switch closes. Switch LE also closes a contact LE_1 in series with leveling slow-down relay coil LZ. Making this coil alive opens contact LZ across a section of the generator leveling shunt-field resistance. This section is short-circuited by contact LE_2 and does not come into effect until the car is leveling into the floors.

Car-passenger Floor Stops. Floor stops for passengers in the car are made by pushing buttons located above the starting switch (Fig. 9-3). The buttons are numbered according to floors at which stops can be made. When they are pressed they are held closed by one large magnet in back of them. The circuit for this magnet when the car is in up direction is from the $+02$ line, through stopping-switch contact $F2U$ on top of the car, P_1 contact on up-direction switch, resistance AR_{35}, and car-button-magnet coil, to the -01 line (Fig. 9-6). This magnet is energized all the time except at the terminal landings.

As the car approaches the top landing, contact $F2U$ opens the circuit through the car-button-magnet coil and the buttons are released. Near the terminal, the down-direction switch coil is energized, closing that switch and opening the up-direction switch. When the down-direction switch closes, it also closes contact P_2 in the car-button-magnet circuit through stopping-switch contact *F2D*. The car-button magnet is then energized to hold the buttons closed when they are pressed for down-direction stops.

As shown in Fig. 9-7, when the car-stop buttons are closed, they complete a circuit to a contact on the selector. For example, if button 14 is pressed, it is held closed by the car stop-button magnet, and it completes a circuit to C_{14} contact on the selector. On the selector carriage are two brushes, CC_1 for up-direction stops and CC_2 for down-direction stops. Brush CC_1 is set a little higher on the brush carriage than is CC_2, so that in up travel CC_1 brush is in advance of CC_2 and in down direction CC_2 will be ahead of CC_1. This is necessary because up stops must be picked up when the car is below floors and down stops when the car is above floors.

Auxiliary contact *P1a* on up-direction switch P_1 is closéd to complete the stopping circuit (Fig. 9-7). Contact *P2a* on the down-direction switch is open and prevents down-direction brush CC_2 from making a circuit for up-direction signals. In down direction, *P1a* contact is open, *P2a* closed, and CC_2 contact makes the stopping circuit.

Closing No. 14 car-stop button when the car is going up, brush CC_1 picks up stationary contact C_{14} on the floor selector and completes a circuit from the $+04$ line, through *OSa* and *HL* contacts, rear coil on *R* switch, *CBR* resistance, contacts R_2 and *P1a*, selector contact C_{14}, and stop button 14, to the -01 line.

Switch *R* is a quick-release type with a light-steel armature held to the contactor arm by two spiral springs. When the switch is closed by the front coil the armature is pulled away from the contactor arm about $\frac{1}{4}$ in. When the rear coil of *R*, which is connected to oppose the front coil and to have greater magnetizing power than ordinarily, is energized, the armature is released and is snapped back by the springs against the contactor arm, which drops quickly to the open position.

When *R* opens, it breaks the stopping circuit by opening contact R_2, and deenergizes the rear coil or switch *R*. Contact *R* in series with the front coil of switch *R* is opened and deenergizes the front coil of this switch so that it cannot close again. Switch *R* cannot close now, even if the car switch is held in the start position, because contact *OS* in series with this switch is open as long as switch

OS is closed. When contact *R* opened, it also broke the circuit through coil *EE* of the auxiliary second-speed switch and the circuit through the floor-selector brush magnet coil *BM*.

Energizing the brush magnet moved the up-direction brushes a short distance in the up direction and the down-direction brushes in a down direction. Deenergizing the brush magnet allows the up-direction brush to move downward and the down brushes upward. This action delays the brushes on the stopping contact to hold the connections closed until switch *R* has opened the circuit and relieved the selector contacts of this service.

When *R* opened, it broke contact R_3 in series with high-speed switch coil *F*. This switch opens its contact F_1 and connects the section of resistance between 4 and 7 in series with the generator shunt field. This reduces the generator voltage, causing the elevator motor to slow down. At the same time F_2 contact, in parallel with motor-field resistance 1FR2, closes and short-circuits this resistance out of the motor field. This increases the motor-field strength and causes it to slow down, as did connecting resistance in series with the generator shunt field.

As previously mentioned, coil *EE* was killed when switch *R* opened, and this in turn opened contact *EE* in series with high-speed switch coils *E* and *F*. Switch *F* opened immediately when *R* opened, but *E* was delayed in opening, having an inductive circuit closed through reactor No. 2 and resistance *EPR*, until after *F* had dropped out. When *E* opens, it puts the resistance section between 3 and 4 in series with the generator shunt field, causing a further slowing down of the elevator motor. Contact *Ea*, in series with the selector pawl magnet coil, is also opened, contact R_1 having been opened previously when contactor *R* opened.

When the pawl magnet is killed, it releases its pawls, which engage a U-bolt on the rods that operate the stopping switches on top of the selector. First, contact *2U* opens and breaks the circuit through high-speed switch *D* coil. When *D* opens, it connects the resistance section between 2 and 3 in series with the generator shunt field, and the motor slows down. Voltage across the motor terminals has been reduced to where contact Q_3 opens and cuts the resistance section between 7 and 8 in series with the generator shunt field. The motor is slowed down now to normal slow speed. Next, *1U* opens and deenergizes coil *Mc* of the microselector device. Up-direction microleveling selector contact *MU* then closes the circuit through up-direction leveling-switch coil *2LA* and leveling-field and brake-switch coil *LH*. Contacts *2LA* close and connect the generator leveling-field winding in circuit, but this circuit is still open through contact 1*Ha*. *LH* coil closes contacts LH_1 in parallel with contacts 2*H* and 1*H*, in the brake-coil circuit, so that the latter two may open and the brake coil remain energized.

An instant after 1*U* selector-switch contact opens, 3*U* breaks and deenergizes coils 2*H* and 1*H*. These switches drop out and open contacts 1*H* and 2*H* in the brake and generator-field circuits, and close contact 1*Ha* in the leveling-field circuit. This kills the running shunt-field winding and makes the leveling shunt-field winding in the generator alive.

When *LH* switch pulled in, it closed contact LH_1 across resistance $1FR_1$ in the elevator-motor field circuit to keep this resistance cut out of the circuit when switch 1*H* opened. Contact *LHa* also closes and completes a circuit for *Y* coil from the $+04$ line, through resistance *YR*, coil *Y*, and contact *LHa*, to -01 line. Coil *Y*, when made alive, opens contact *Y* and puts the generator series-field winding in circuit for leveling. Contact 1*Ha* opens in *OS* coil circuit and this switch drops its several contacts out.

Magnet coils *O* and *BM* were deenergized when the running shunt-field winding on the generator was opened. Contact O_1 in parallel with the car-switch contact opens and kills right-coil *MM* of the idling-switch relay. This relay is held closed, however, by coil *MM* connected across the elevator-motor terminals, until the motor comes almost to rest. Killing *BM* coil opens contact *BM* in the pawl magnet circuit, which has previously been killed by the opening of contact *R*.

Microleveling Operation. As previously described, when the coil of the microselector device is deenergized, control of opening and closing of the microcontacts is taken by a cam. At first, this cam prevents the high-speed leveling contacts from opening. As the cam continues to turn with the motion of the car, contacts *H* are allowed to open and break *LE* coil circuit. Killing this coil opens contact LE_2. Opening this contact groups in parallel resistances 4*FR*1 and 4*FR*2; these two, in series with resistance 4*FR*, are now in series with the leveling shunt-field winding. Weakening the generator field causes the motor to slow down.

When *LE* opened, it also broke the circuit for leveling fast-speed switch coil *LZ*. Opening of this switch is delayed by coil *LZ* being connected in the inductive circuit formed by reactor No. 3 and resistance *LZR*. When *LZ* coil releases its armature, it closes contact *LZ* to shunt resistance 4*FR*1 and 4*FR*2 out of the leveling field-winding circuit. At about the same time, the microleveling cam releases contact *MU* and breaks the circuit through 2*LA* and *LH* coils. These contactors drop out, opening the brake-coil circuit, and the brake is applied to stop the car level with the floor. LH_1 contact opens and cuts resistance 1*FR*1 in series with the elevator-motor field winding. Contacts 2*LA* open to disconnect the generator leveling shunt-field winding; and contact 2*LAa*, in series with resistance 4*FR*2, breaks the circuit through this resistance.

When the car stops level with the floor, the rolls in the two arms *A* that control opening and closing of the microleveling contacts just clear their leveling cam (Fig. 9-15). If for any reason the car stops 0.5 in. or more above the floor, the down slow-speed leveling contact *MD* is closed by the leveling cam. This energizes the brake and the slow-speed leveling circuits to bring the car back level with the floor. As the car comes to rest at the floor, all contactors are in normal running position.

Door-opening Operation. When the car is leveling into the floor, car and hoistway doors are being operated so that they will be open at about the same time that the car levels with the floor. When first-speed magnet *D* opens, it closes contact *Da* in the door-operator circuit. Contact *PP* in the door-opening circuit closes when selector contact 4*U* opens the circuit through auxiliary door-open relay coil *PP*. Contact *LHa* closes when the microleveling circuits are energized and contact LE_2 closes when the high-speed microleveling circuit is deenergized.

When contact LE_2 closes, it completes a circuit for the front coil on door-operating relay *P* from +02, through front coil *P*, resistance PR_1, contacts LE_2 and *LHa*, reactor No. 1 and PR_4 resistance in parallel, and contacts *PP* and *Da*, to the —01 line. Contacts *P* and *P* close and complete the circuit for the door-opening switch *DO*. This circuit is from the +04 line, through the limit switch Lt_2 (closed when the doors closed), resistance *DOR*, coil *DO*, stall-relay contact 04, and contacts *P* and *P*, to the —01 line. Coil *DO* closes direction contacts *DO* and *DO* and opens contact DO_1. This energizes the motor to open the doors, operation being much the same as when they were being closed. When the doors have opened, limit switch Lt_2 opens and stops the motor, and the brake is applied to hold the doors in the open position.

Microemergency Leveling. The microemergency-leveling switch may be used to bring the car to the floor in case it cannot be operated from the start switch. With the hoistway doors and car gate closed, the door-contact relay functions and closes contact G_1. Car gate No. 1 contact also is closed. These make the microemergency switch alive. Contact *Ra* is also closed because contactor *R* is open. If the microemergency switch is closed to the down position, its right-hand pole makes the high-speed leveling switch coil *LE* alive. Its left-hand pole energizes the up-direction coil *2LA* and the car starts. Throwing the microemergency-operating switch to the up position will energize the down-direction leveling switches, and the car will move downward. When operated from the micro-emergency switch, the car runs at fast microleveling speed, which is about 50 fpm.

Stops Made from Hall Buttons. Stops made from hall buttons being pressed by waiting passengers involve a somewhat more complicated selector and relay arrangement than do those made from car buttons. In the car one set of buttons serves for both up and down directions and one magnet coil serves to hold the floor buttons in the closed position. At each landing, with exception of the terminals, two call buttons are provided, one for up and the other for down direction. When the buttons are pressed the calls are registered by relays, one relay for each button. On these relays are two coils. One of the coils is energized when its hall button is pressed and closes a contact. This contact makes a stopping circuit to a stationary contact, on the selector, to be picked up by a brush on the selector carriage. The second coil on the relay is energized by another contact on the selector, to reset the relay when the floor call has been made. The hall push-button relays are assembled on a panel, as shown in Fig. 9-11.

Provisions are made to bypass a floor call to a following car. This is done by relay *NS*, the coil of which is in series with the non-stop buttons, one in the car and the other on the dispatcher's board. These two buttons are normally closed and make a circuit for *NS* relay coil from $+04$, through *NSR* resistance and coil *NS*, to -01 line. This relay closes a contact *NS* in the common lead to down-stop brush *CD'* and up-stop brush *CU'* on the selector carriage.

Contact *NS* is normally closed and the selector contact circuit is complete for the cars to pick up floor calls. If the car is full and the operator does not wish to pick up any more passengers, by pressing the nonstop button he causes the *NS* relay contact to open to prevent the floor calls from being picked up. They will be bypassed to the following car.

P1a' and *P2a'* contacts in series with *CD'* and *CU'* selector brushes, respectively, allow the car to be stopped for calls corresponding to the direction in which it is going. *P1a'* contact is open when P_1 up-direction switch closes. This breaks the circuit to down-direction-call selector brush *CD'*, and it cannot stop the car when it is going up. In down direction $P1a'$ is closed and $P2a'$ open.

The reset brushes SL_4 and SL_1 on the selector are also in series with $P1a'$ and $P2a'$ contacts on the direction switches, so that only up-direction calls will be reset when the car is in up motion, and down-direction calls when the car is in down motion. The $1Ha$ contact in series with both selector brushes SL_4 and SL_1 is open as long as switch $1H$ is closed, when the car is operating above micro-leveling speed. Therefore, unless the car stops at a floor, the floor-call signals will not be reset.

Fɪɢ. 9-11. Hall-call push-button relays are assembled on this panel.

The hall lights, which indicate to waiting passengers when and in what direction a car is approaching, are also controlled from the selector. Stationary *FU* contacts on the selector are for up-direction lights and *FD* stationary contacts are for down-direction lights.

When the car is going up, selector brush SL_1' is connected into the circuit by the P_4 relay contact, $P4a$ being closed when the relay is killed by contact $1P_1$ being open when up-direction switch P_1 is closed. In down direction, P_1 direction switch is open and $1P_1$ contact, in series with P_4 relay coil, is closed. This makes coil P_4 alive, and contact $P4a$, in series with selector brush SL_4', closes so that the down-direction lamps will light. Contact $P4a$, in series with selector brush SL_1', is open; therefore the up-direction lamps cannot light when the car is going down.

The current to light the hall lamps is controlled by relay *HL*, the coil of which is in series with contact $1D$ on top of the selector. When the pawl magnet releases the contacts on top of the selector, and the car starts, $1D$ closes and completes the circuit for relay coil HL from the $+02$ to the -01 line. Contact HL_1 opens and breaks the power circuit to the hall lamps. The Ca contacts in the hall-light circuit are on the potential switch and close when this switch pulls in, so that as long as this switch is closed, power is on the hall-light circuit.

Contact NS_1, in series with the $+LS$ line, leading to the common connection for the hall lights, is on the nonstop relay and is normally closed. If the nonstop switch is open in the car or on the dispatcher's board, for the car to run nonstop, NS_1 would have to open and cut power off the hall lamps. This would prevent their lighting even if the opposite side of the lamps were grounded.

The nonstop relay has a third contact NS_2, in series with a light on the dispatcher's board. This contact closes when the NS relay is dropped out by the opening of a nonstop switch, and it lights the nonstop lamp on the dispatcher's board. The chief purpose of this light is to show the dispatcher when the operator is running nonstop.

Dispatcher's Board. Lights on the dispatcher's board show the direction in which the car is going. These lights are controlled by relay P_3. This relay has two coils. One of them, the front coil, is connected directly from the $+02$ to the -01 lines and is energized all the time that the stop switch in the car is closed. The other P_3 relay coil is connected across the elevator motor, so that when the motor reverses, the direction of current flow in this coil is changed. The two coils of relay P_3 are on the same core; consequently, with the motor running in one direction the coils assist each other. When the motor runs in the opposite direction, the two coils on P_3 have opposite polarity.

Front coil P_3 cannot close the relay alone; therefore, as long as the rear coil opposes it—such as when the elevator is in the down direction—the relay will not pull in, even when the elevator is stopped. Under this condition down contact P_3 will remain closed and up contact P_3' will stay open. When the car starts up, the polarity of the motor reverses, the two P_3 relay coils have the same polarity, and the relay opens. When the relay closes, its front coil can hold it closed, keeping down contact P_3 open and up contact P_3' closed. Thus contacts P_3 and P_3' are controlled automatically with the direction of the motor.

Car-position Indicator. On the selector are shown contacts for the car-position indicator, which shows car passengers the number of the floor that the car is approaching. There are a single row of stationary contacts on the selector for this device and two groups of two brushes each on the carriage to operate the position indicator. The up- and down-direction circuits are controlled by relay P_4. In the up direction, this relay is deenergized and contact P_4, in series with the up coil of the indicator, is closed. In the down direction, relay P_4 is energized, opening contact P_4 in the up-direction coil circuit and closing contact P_4' to complete the down-direction coil circuit of the indicator. The coils turn an indicator to show floor numbers, which are illuminated by light L, connected from the $+04$ to the -01 lines.

Assume that the car is going in the up direction and approaching the fourteenth floor. Brush CU_1 on the selector picks up the stationary contact on the selector and energizes the up coil in the indicator, and the number 14 is flashed into view. Before CU_1 moves

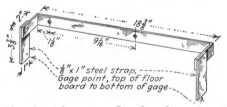

FIG. 9-12.   Gage for setting floor boards on the selector.

off the stationary contact, brush CU moves on and retains the circuit through resistance EMR_2. This is done to reduce the current in the circuit before it is broken, to prevent severe arcing at the contacts. In the down direction, brushes CD_1 and CD close the down-direction coil circuit in the indicator.

Floor-call Relays.   At the top of Fig. 9-7 are shown floor-call push buttons in the hall- and floor-relay magnet coils. As previously mentioned, each relay has a magnet coil and a reset coil. For example, magnet coil U_{14} and reset coil US_{14} are parts of the same relay. Assume that up-direction button U_{14} on the fourteenth floor is pressed. This completes a circuit for relay magnet coil U_{14} from the $+LS$ line, through the coil and button U_{14}, to the $-LS$ line. Energizing this coil closes contact 14, shown between relay reset coil US_{14} and selector stationary contact US_{14}'.

When selector brush CU' touches contact US_{14}', a circuit is completed for the rear coil of switch R, from the $+04$ line, through contacts OSa and HL, coil R, contact NS, resistance RR_1, contacts $P2a'$ and CU' reset coil US_{14}, to the $-$ line. Making rear coil on R alive initiates the stopping operations as explained for stopping from the car buttons.

At about the time that brush CU' picks up contact US_{14}', brush SL_1' picks up contact $14FU$. Then, as soon as the pawl magnet is released and selector contact $1D$ is opened to kill hall-light relay HL, contact HL' closes and the up-direction lamp on the fourteenth floor lights.

The current through reset coil US_{14} with coil R in series is not

Fig. 9-13. Bottom of floor selector, showing how the vertical U-bolt rods R and R_1 are connected by tie rod T.

sufficient to reset the relay. As the car approaches the floor, up reset brush SL_1 picks up selector stationary contact UR_{14} to complete the US_{14} reset-coil circuit, as soon as $1H$ relay opens and closes contact $1Ha$. The reset circuit is from the $+04$ line, through the two nonstop buttons, resistance AR_{34}, contacts $1Ha$, $P2a$, SL_1, and US_{14}', floor contact 14, coil US_{14}, to the $-$ line. Making coil US_{14} alive resets the relay, and contact 14 opens to cut out the floor call. Floor calls from other buttons are taken care of in the manner just explained for the fourteenth floor, but with the selector contacts corresponding to the floors and the direction of the car.

Adjusting the Selector. When the elevator is installed, the floor boards on the selector are properly adjusted and should not require further attention. This adjustment is made by placing the car platform level with the top or the bottom landing. Then, with the selector disconnected from the tape-sheave shaft, its crosshead is adjusted to where the floor boards will have to be moved a minimum distance to bring them to a correct position. A floor gage is fastened to the crosshead, and each floor board is adjusted so that it is in exact line with the gage when the car is level with a corresponding floor.

Figure 9-12 shows the dimensions for a floor gage for a type 11*SLU* controller, mounted by bolts G on the brush panel, as shown

Fig. 9-14. Pawl magnet P on floor-selector crosshead.

in Fig. 9-13. The gage for other types of control may vary slightly from the one shown. If a gage is not at hand, it is well to obtain one from the manufacturers in case checking the position of some of the floor boards should be required.

The normal or release position of the U-bolt rods is adjusted by means of a bumper B at the bottom of the selector, as shown in Figs. 9-4 and 9-13, so that the bell crank which connects to the up rod R moves an equal distance above and below a horizontal center line when its pivot is raised $13/16$ in. The tie rod T between the two vertical rods is used to accomplish the same adjustment on rod R_1 as given for down rod R. The bumper and tie-rod adjustments are made at the factory and under normal conditions should not require attention after the equipment is in place.

At the time the floor boards are adjusted, the U-bolts should be set. This is done by raising up U-bolt rod R $1\frac{3}{16}$ in. and securing the rod in that position. With the rods blocked in position and the car level with the landing the U-bolt on the up rod should be adjusted so that the up pawl just slips under it. On the down rod the U-bolt is adjusted so that the down pawl will just pass over it. There should be no clearance between the U-bolts and the pawls, and the U-bolts should be locked securely in place. The pawls are shown engaging the U-bolts at C in Fig. 9-14. Engagement of the pawls with the U-bolts is adjusted by bolts B. This adjustment

FIG. 9-15.   Brush magnet N on floor-selector crosshead.

should be made to allow the pawls to engage the U-bolts with a firm contact.

When the floor boards and U-bolts are adjusted, set the cams for the microleveling control. When the car is level with the floor, the leveling cam is set so that it will be in line with, and will clear by equal distances, the rolls in the two arms on the selector crosshead. In Fig. 9-15 cam L_1 is shown in the correct position with relation to the rollers in arms A when the car is level with a floor landing.

After final setting of floor boards and U-bolts, the car should be placed level with the top floor and the position of the crosshead guide arms marked on the square vertical bars. This marking is necessary and will be found convenient should the tape or chain break on the selector drive. To obtain approximate setting of the selector, the crosshead is run by hand to the marks on the square bars and the car set at the top floor. The selector is then connected to the drive and a test made to see how the car levels at the floors. It may be necessary to make changes in the drive to bring the car level with the floors. If the adjustment is small, it can be made by shifting the selector's sprocket wheel on its hub. When this is not sufficient the chain may be opened and the crosshead moved by the sprocket wheel.

The stopping contacts at the top of the selector are adjusted by adjusting screw *A*, shown on the front row of contacts in Fig. 9-16. Similar adjusting screws are on the other row of contacts. Contacts

Fig. 9-16. Top of the floor selector showing stopping contacts on the right and microleveling contacts on the left.

1*U*, 2*U*, and 3*U* are adjusted for the up motion of the car, and although these settings vary somewhat for different car speeds, the following are about correct for car speeds of 700 to 800 fpm: Contact 1*U* is adjusted to open when the car is 26 in. below the floor, 3*U* to open when it is 12 in. below, and 2*U* to open when it is 5 ft below. Contact 1*D* may be adjusted with the car moving in either direction and is set to open when the car is within 8 in. of the floor.

Contacts 2*D* and 3*D* are set with the car in the down direction. They are adjusted so that 2*D* opens when the car is within 5 ft of the floor and 3*D* when the car is 12 in. above the floor. Contacts 4*U* and 4*D* control the opening of the doors and are adjusted to cause the doors to open at the desired time.

Opening of the microleveling contacts on the selector is timed by the adjusting screws *F* on the high-speed contacts and *S* on the slow-speed contacts. The fast-speed contact *H* is adjusted to open when the car is 5.5 in. above the floor and contact H_1 to open when the car is 5.5 in. below the floor. In each case the car is moving toward the floor. The two slow-speed microcontacts *MU* and *MD* are adjusted to open when the car is within ¾ in. of the floor. Contact *MD* is for down motion and *MU* is for up motion of the car.

Distance of Car from Floors When Hall-stopping Contacts on Brush Magnet Engage Floor-board Contacts, Magnet Energized

Car speed, fpm	Up travel, ft in.		Down travel, ft in.		Car speed, fpm	Up travel, ft in.		Down travel, ft in.	
800	17	0	18	0	600	10	8	11	4
750	15	4	16	0	550	9	8	10	0
700	13	7	14	4	500	8	8	9	0
650	12	0	12	8					

The floor-stop contacts on the selector crosshead are carried on the brush magnet. There are two sets of these contacts, one for up motion and one for down motion, and each is carried on an adjustable

Fig. 9-17. Selector brush-magnet and floor-board contacts.

armature on the brush magnet *N*, shown in Fig. 9-4. The general arrangement of the brush-magnet and stopping contacts is shown in Fig. 9-17. Stopping operations in floor-to-floor travel are adjusted by screws S and S_1 for down motion. Similar adjustments are made on the top armature for up motion. These adjustments are independent of others and will not be affected by them. The stroke of the magnet is adjusted by screws S_2, so that the car-stopping contacts just make the corresponding floor-board contacts when the car is a distance away from the floor equal to that shown in the table.

The contacts shown at *C* in Fig. 9-13 for illuminating the hall lanterns, the car-position indicator, and other signal devices are mounted on an insulating board. This board is slotted so that the contacts may be moved vertically to allow adjustment of each signal device to function at the proper time.

Judgment must be used in applying the instructions given in the foregoing. The settings suggested are for one type of control; adjustments for other types may vary slightly from those given. If the instructions are intelligently applied, they will be of great assistance to elevator mechanics who have not had an opportunity to become entirely familiar with this equipment.

CHAPTER 10

Automatic Dispatching of Passenger Elevators
and Attendantless Operation

Elevator-service Needs. Regardless of the type of building, elevator traffic varies widely at different hours of the day. Therefore the dispatcher has difficulty regulating operations to give equal service to all floors. To overcome this long-recognized problem, elevator engineers have developed several signal systems that help the dispatcher render better service from a given number of elevators. Comparatively recently their efforts have produced supervisory-control systems that automatically dispatch a bank of elevators according to traffic needs.

Service demands on a bank of elevators occur in waves rather than in a steady flow (Fig. 10-1). With variations, these demands can be classified as: (1) up-peak period, when traffic is largely incoming, as in the morning when people come to work; (2) off-peak demand, a long period of two-way traffic with several up-and-down peaks, some relatively severe; (3) down-peak conditions, when traffic is mostly down as people hurry out of the building at closing time.

Supervisory System. One of the supervisory systems which make it possible to operate a bank of elevators at high efficiency for every traffic condition is known as the Selectomatic, by Westinghouse. It can be applied to any bank having two or more passenger elevators.

To put into effect the proper system of supervisory control for any traffic condition, the dispatcher simply presses a button. Pressing one button brings elevators into operation to handle up-peak service. Another button schedules elevators to serve down-peak traffic. If neither button is in use, elevators function for two-way traffic, including peaks.

Dispatcher's Panel. At the starter's station on the ground floor is a panel (Fig. 10-2) through which the several supervisory operations are initiated. At the top center of the panel are columns of illuminated numbers that show the position of each car in the hoistway. The column on each side shows the registered up-and-down corridor calls. Just below the car-position lights and running from right to left are rows of lights indicating travel direction of each of six cars, other supervisory signals, and the bypass light for each car.

Just below these lights is the *SM* button which, when pressed, puts the supervisory system in operation, and the *Q* button, which increases by one the established corridor-call quota of the low-rise cars.

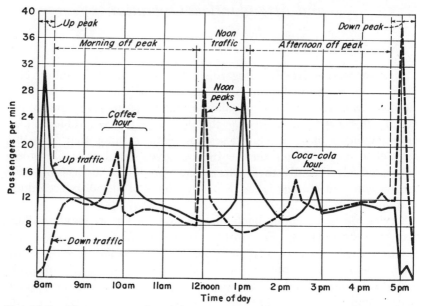

Fig. 10-1. These curves, plotted from a traffic study made in an office building, give a general idea of the peaks that occur in passenger travel on modern elevators.

Between these two buttons is a knob to increase or decrease the established dispatching interval between cars at terminal floors as they are put into or taken out of service. *U* button puts cars on up-peak operation, and *D* button initiates down-peak service. Below are four buttons to select cars for low-zone service.

The in-service buttons put cars in or take them out of service. Below these buttons is a row of call-backs with which the starter can call any car back to the bottom landing. Then comes a row of bypass buttons with which the starter can cause a car to bypass calls, and let a following car answer them. The keyed motor-generator switches for each elevator and the lights which show if these machines are in service are at the bottom of the panel.

Supervisory-system Operation. How does the system supervise a bank of elevators in up-peak service? This is done in either of two ways, depending on the rate at which people enter the building. If there are enough cars to handle the traffic demand adequately in the morning—and there generally are—the starter presses up-peak button *U*. Each car travels to the floor of the highest call. Then, when the operator puts the car switch in start position, the car immediately reverses and returns to the ground floor.

All cars are dispatched up automatically from the ground floor at regular intervals, except that, if they are full, they can be signaled away by the starter. The next car will be dispatched a full time interval from the time the previous car leaves the terminal. Cars are dispatched down from the floor of the highest call on an instant start-down signal. All floors are served by all six cars and all corridor calls are common to all cars.

Zone Up-peak Operation. When up-traffic flow is extra heavy and cannot be adequately handled by all cars going to the highest call, which seldom happens, elevator capacity can be increased by changing to zone operation. To do this the elevators are divided into two groups and the building into two zones. One group of cars, say 1, 2, and 3, serves the high zone, while the other group, 4, 5, and 6, serves the low zone.

The starter initiates this operation by pressing the low-zone buttons for the cars he wishes to serve this zone. Passengers are directed to the cars serving their floors by lighted signs at the first-floor terminal. The cars in each group are dispatched up automatically from the ground floor on regular time signals, independent of those in the other group.

Off-peak Operation. During most of the day, when up-and-down traffic is about equal, all cars operate on a through-trip basis under automatically timed dispatching from top and bottom terminals (Fig. 10-3). Thus, cars are signaled in sequence at regular intervals from the terminal landings so that all floors get uniform service. The starter puts elevators in off-peak operation by pulling out on his panel the peak-up and low-zone buttons if any are pressed in.

When the number of calls in each direction is uniform, the cars remain fairly evenly spaced. But there are short, heavy demands, as when meetings gather or close, or when it is lunch time. During these traffic peaks, car round-trip time increases and the scheduled dispatching interval is too short. Cars cannot leave on schedule and there are long waits at some floors. When traffic becomes light, the round-trip time of each car decreases and the dispatching interval is too long.　Cars then arrive ahead of schedule and bunch at terminals.

Cars behind Schedule.　To decrease the round-trip time of a car behind schedule, the supervisory-control automatically reverses the car as soon as it answers the highest call.　This occurs if there is no car at the top floor when the start-down signal is given.

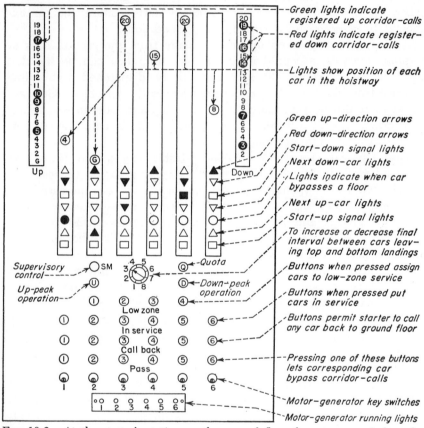

FIG. 10-2.　At the starter's station on the ground floor there is a panel through which the several supervisory operations of the passenger-elevator bank are initiated.

To decrease the dispatching interval and avoid bunching the cars at the terminals, the antibunching feature comes into operation.　If a car is waiting at the terminal when another car arrives, the interval of the first car is decreased one-half.　If a third car arrives, the regular interval of the first car is decreased to one-quarter.　If a fourth car arrives, the first car's regular intervals is reduced to zero and it is dispatched immediately from the terminal landing.

When down traffic is heavier than up, cars tend to arrive at the top terminal ahead of schedule and so are forced into down traffic earlier than normally. This increases the time available for each car to make the down trip and to handle traffic increase without exceeding the normal round-trip time. If the heavier traffic is in the up direction, the system handles the increased up load in a similar way.

Down-peak Operation. Down-peak conditions usually occur in office buildings for a relatively brief period around 5 P.M., when practically all the building's population must be brought to the ground in a short time. At the start of the period the dispatcher pushes button D (Fig. 10-2). Then the supervisory system automatically divides the cars into two groups similar to those for up-peak operation. One group of cars serves the high zone or upper floors, while the other group serves the lower floors. Each group has its own timing dispatcher, which operates independently.

High-zone Cars. Cars assigned to the high zone handle all traffic in the high zone and all up calls in both zones. The reason for the latter is that up passengers are taken directly to their floors. If a low-zone car stopped for an up passenger who was going to the seventeenth floor, the passenger would have to transfer at the fourteenth. By having the high-zone cars handle all up traffic, passengers are sure of direct service. Usually at this time high-zone cars have only a few up calls. Therefore this traffic has little effect on the high-zone cars' operation.

When this phase of supervisory control is in operation the high-zone cars bypass all down calls in the low zone, but if a down call in the low zone has been bypassed by a low-zone car because it is filled to capacity, the first unfilled high-zone down car automatically stops for down calls registered in the low zone during this down trip only.

Low-zone Cars. Low-zone cars answer all low-zone down calls but no up calls. A car in the low-zone group travels to the highest down call registered in the low zone provided enough calls have been registered to fill an assigned quota. This quota represents the average expected number of stops a car is required to make. It is based on a traffic survey of the building population.

When the number of calls registered in the low zone is not enough to fill the quota of a low-zone car, it goes to the highest call in the high zone, and stops to pick up passengers. Then it reverses and answers its quota of down calls to help out the high-zone cars.

At the moment a quota number of calls is registered in the low zone, the low-zone car that has traveled into the high zones makes a useful stop at the next down call. It immediately returns to the low zone where the traffic demand is concentrated, bypassing all other down calls that may have registered in the high zone.

Thus, the quota is only a measuring stick to determine if a low-zone car will enter the high zone to help out there. But if there are

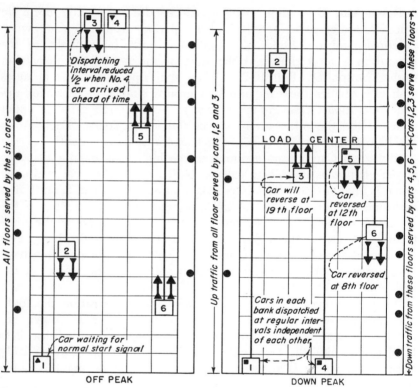

FIG. 10-3. During most of the day, when up and down traffic is about equal, cars run through trips, dispatched automatically.

FIG. 10-4. At the beginning of the down-peak period, the starter pushes a button that divides the cars into two groups.

no calls registered above the load center, the low-zone car reverses automatically at the load center and returns to the ground floor, answering any call registered.

Quota Number of Calls. Stop quota of the low-zone cars is filled when an assigned number of calls has been registered in the low zone. This quota may be set to meet the traffic condition in the building at any given time or season, and may be 2, 3, 4, or more stops. Regardless of the number selected for the quota when the elevators were installed, the dispatcher can increase it one by pressing a button. Since all down calls in the low zone are common to cars in the zone, the number of down calls required for a car to receive its quota depends on the number of low-zone cars going down at a given time.

Let's say three low-zone down calls are the quota for cars in that zone and a car starts down. If no other cars are traveling down, any three calls below the load center will fill the down car's quota. When two cars are going down in the low zone at a given instant, six down calls must be registered to fill the second car's quota.

Consider the typical down-peak traffic conditions in Fig. 10-4. There is a fairly even distribution of down-traffic demand between the high and low zones. Cars 1, 2, and 3 are assigned to the high zone and handle the up traffic during down-peak operation, as well as down traffic in the high zone. Cars are dispatched in both zones at regular intervals independently of each other.

Car No. 3 in the high zone rises to the nineteenth floor, where the highest call is. There it reverses and starts down to pick up calls in the high zone. Car 5 in the low zone rises only to the twelfth floor for its highest call, since in this case its quota is filled, six calls for it and car 6. At the twelfth floor, car 5 reverses and starts down to answer the calls. The system adjusts quickly to handle any other traffic conditions.

This is a general picture of how this automatic supervisory system operates to give uniform elevator service to all floors in a building. It permits the dispatcher to change easily and quickly to operating procedure to meet almost any emergency. The system's flexibility stems from the selection of the load center of traffic demand in the building, the number of cars that may be assigned to each zone, and the passenger-call quota selected for the cars.

Another system for dispatching a bank of passenger elevators automatically is by Otis and is known as Autotronic. This system, like Selectomatic, can do all, and far more than, an expert supervisor can do manually even if he gives his entire attention to dispatching a bank of elevators. These controls take care of so many of the supervision details automatically that the dispatcher can give a large part of his time to the needs of incoming and outgoing passengers.

Indicator Unit. A supervisory indicator unit (Fig. 10-6) forms part of the Autotronic system. On the left in its top section is a column of up-direction green numerals, and on the right a column of down-direction red numerals. These numerals light when corridor buttons are pressed to show the supervisor where passengers are waiting on floors. When a car answers a call the corresponding light goes out. Between these two columns are amber landing numerals. As a car approaches a floor the corresponding numeral lights; it then goes out as the car leaves.

Below these numerals are two rows of arrows, the top ones green for up-car travel and the bottom ones red for down. When a car

FIG. 10-5. Electronic supervisory control panel from which elevators are dispatched in the 100 Park Avenue building, New York.

is making an up trip its green arrows light and when it is coming down its red arrows light.

Below these arrows a row of numerals remains dark except when a car operator holds down the nonstop switch, which usually means that the car is full and is going express to the main floor.

Dispatching Lights. At the bottom on the left of the supervisory indicator (Fig. 10-6) is the green up group-dispatching light *U*. When lighted it indicates that the next car to arrive at the bottom floor will get an up dispatching signal immediately. The red group-dispatching light *D* on the right, when lighted, shows that the next up car to arrive at the top landing will be immediately dispatched downward.

Between the U and D lights are two rows, top green for up direction, and bottom red for down. These are the car-dispatching lights. When one lights it indicates that a corresponding car has been dispatched away from a terminal landing. If a green lamp lights, the car should start in the up direction, but when the red lamp lights, a corresponding car should start down.

Signal lights on this panel give the elevator supervisor at all times a picture of where each car is, its direction of travel, how it is progressing toward a terminal landing, the floors at which passengers are waiting, and other pertinent information. Knowing all this, the supervisor needs a means to convert this knowledge automatically into efficient elevator operation, which he has with the supervisor's control unit (Fig. 10-7).

Supervisory Unit. This is usually mounted below the indicator unit to form an indicator and control panel (Fig. 10-5). Switches on the supervisory unit are mounted on a panel tilted back and illuminated from above for the supervisor's convenience. Center-opening laminated glass swing doors with a magnetic catch guard against unauthorized tampering.

A dispatching-interval switch, shown enlarged in Fig. 10-9A, sets the dispatching-time interval between cars signaled away from the terminal landings. When the switch is set on 1, a car is dispatched every 15 sec from the bottom or the top landing. Setting the switch on 7 increases the dispatching-time interval to 50 sec. Any interval between these two can be had by positioning the switch.

On the right-hand side of the panel shown in Fig. 10-7 is the traffic-flow dial switch, enlarged in Fig. 10-8. Each setting of this switch represents a specific program of operation, one program for each of the six types of traffic. Moving the switch from one setting to another puts a new program into operation and cancels the previous one. Fig. 10-10 shows the six programs, their dial settings, and the conditions under which they are used.

Of the four traffic-flow lights (Fig. 10-9B) two green arrows point up and two red arrows down, indicating which traffic program is operating. Lights are numbered for easy identification. The ones that are lighted for the different services are indicated in Fig. 10-10. For example, during up-peak service, green lamp No. 2 lights, and for balanced service, green lamp 2 and red 3 light.

Dispatching Buttons. Manual dispatching buttons U and D (Fig. 10-7) are used to signal a car away from a terminal landing ahead

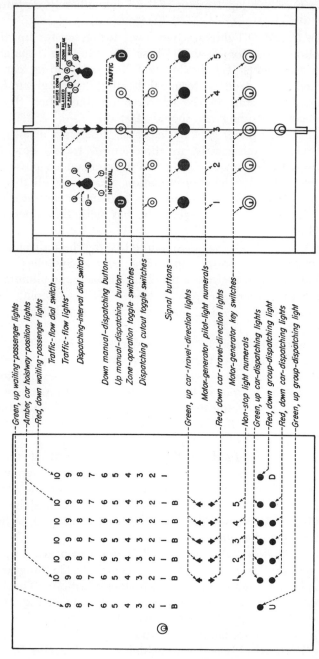

FIGS. 10-6 (left). Supervisory indicator is usually mounted above the control unit, as shown in Fig. 10-5, to form an indicator-control panel. FIG. 10-7 (right). Control unit provides the elevator supervisor with facilities for regulating elevator dispatching as the traffic changes.

of the automatic timing schedule. Pressing *U* button dispatches a waiting car from the ground floor. Pressing *D* button does the same for a waiting car at the top landing.

Zone toggle switches, one for each of the three cars that can operate in low-zone service, are used to transfer cars from low-zone to high-zone operation, or vice versa. When a switch for an elevator

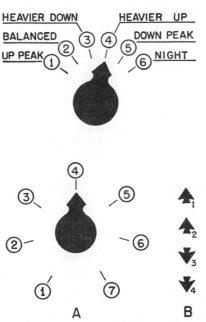

Fig. 10-8 (above). Each setting on traffic-flow switch initiates an elevator-operation program. Fig. 10-9 (below). A, elevator dispatching-time interval switch. B, elevator traffic-flow lights.

is in one position and the system is operating zone return, this car will serve low-zone traffic. When in the other position the elevator serves the high zone.

Dispatching cutout switches, a toggle switch for each car (Fig. 10-7), are used mainly to take cars out of or return them into service. However, if passengers at the low terminal landing enter a car not scheduled for the next up signal, the supervisor may transfer the next up signal to this car. This is done by temporarily turning the cutout switch of the preceding car to the off position.

Below the cutout switches is a row of signal buttons, one for each elevator. These buttons sound a buzzer in a corresponding car for the supervisor to communicate directly with the operator, as by a signal code. When telephones are in the cars, the buttons can be used to tell the operator he is being called.

At the bottom of the panel is a row of motor-generator start-and-stop key switches. Above these are the motor-generator pilot-light numerals that show which of these units is running. Also, if a car is parked at the ground floor with its door closed, turning its motor-generator switch to start opens the doors after the unit comes to speed.

DIAL SETTING	TYPE OF TRAFFIC HANDLED	DIAL SETTING	TYPE OF TRAFFIC HANDLED
UP PEAK (1) *Traffic-flow lamps lighted*	*Heavy – up traffic from main floor with little or no interfloor or down traffic as during morning peak.* *No. 2 green*	HEAVIER UP (4) *Traffic-flow lamps lighted*	*Heavy –up traffic plus appreciable down traffic, as during the late noon peak.* *No.1 and 2 green No. 3 red*
BALANCED (2) *Traffic-flow lamps lighted*	*Traffic about equal in up and down directions.* *No. 2 green No. 3 red*	DOWN PEAK (5) *Traffic-flow lamps lighted*	*Heavy down traffic with little or no interfloor or up traffic, as during the evening peak.* *No. 3 red*
HEAVIER DOWN (3) *Traffic-flow lamps lighted*	*Heavy down traffic plus appreciable up traffic, as during the early noon peak.* *No. 2 green No. 3 and 4 red*	NIGHT (6)	*Light and intermittent traffic, as during nights, Sundays and Holidays.*

Fig. 10-10.   Six elevator-operating programs obtained from setting of the traffic-flow switch shown in Fig. 10-8, conditions for which the programs are used, and the indicator lights.

Up-peak Operation.   At the beginning of the morning up peak the supervisor sets the traffic-flow dial to *Up Peak*. Cars are then dispatched from the low terminal only. They waste no time in useless travel to the top terminal. When they have answered the highest up call, they reverse automatically and return to the lobby.

Car operators are generally instructed to leave the main floor when a car is 80 per cent full. Thus, cars that fill quickly are not delayed. A car that loads slowly does not wait indefinitely for a full load. It is dispatched at a given time after the preceding one, provided another car is ready to load. If a car is not available, the partially loaded one is held for half an interval after its normal leaving time.   Normally there is at least one car ready for passengers at the main landing.

Balanced Traffic.   During balanced up-and-down traffic the chief problem is to keep the cars evenly distributed in the hoistways so that all floors will get regular service.   For balanced-traffic operation the supervisor sets the traffic-flow dial to *Balanced* and the cars go into operation on this program.   Cars are automatically dispatched from both terminal landings at regular intervals in the order they arrive.   But a delayed car can bypass others without upsetting schedules.

A delayed car is forced to make up lost time.   It is reversed before it reaches the top landing, if there are no unanswered calls above. A car arriving late at the main landing may have its loading time reduced to keep it on schedule.

If a surge of traffic occurs and a car is filled quickly at a terminal landing it can leave ahead of time when loaded.   Another car is dispatched ahead of time from the other terminal landing to take its place.   Also, if a car is delayed at one terminal beyond one-half interval, a car is held back at the other terminal.   This prevents bunching the cars at terminals and helps keep them evenly spaced in the hoistway to keep service standards high.

Heavier-down Program.   During early noon, traffic is heavier down with considerable up traffic.   This also occurs in most buildings, but to a lesser degree, during the midmorning and midafternoon "coffee periods."

Because down cars make more stops and handle more traffic during these periods, the down trip takes longer than the up.   To take care of this condition the supervisor turns the traffic-flow switch to *Heavier Down*.   The cars are then dispatched to have more serving down traffic than up.

Heavier-up Traffic.   Following the heavier-down peak there is a heavier-up peak to get people back to their floors.   When this occurs the supervisor sets the traffic-flow dial switch to *Heavier Up*. Then cars are dispatched to keep an extra one in the up direction. When these peaks have passed, the supervisor sets the traffic-flow switch to *Balanced,* and cars are again dispatched to have an equal number going in both directions.

Down-peak Traffic.   During the evening down peak, traffic accumulates at intermediate floors.   To handle this condition the supervisor sets the traffic-flow switch to *Down Peak*.   Cars are then dispatched from the top floor, or some floor near the top, on an electronically timed schedule.   At the main landing, cars discharge their passengers and immediately start back to the upper terminal.   If there are no calls above a late car, it may be reversed before it reaches the top floor to serve the lower ones.

When traffic becomes very heavy, cars may be filled at the upper floors and bypass the lower ones.   But lower-floor passengers do not have to wait indefinitely.   Their total waiting time is added electronically and when it reaches a set limit the control does something to correct the condition.

If two or more lower-floor calls approach the time limit, the control system switches the cars to zone-return operations.   With zone return the building is divided into a high and a low zone and the elevators into groups, one to serve each zone.   Low- and high-zone cars reverse at highest call in their zones.   To help low-zone cars care for low-zone congestion, unfilled high-zone cars answer low-zone calls until filled.   When the peak has been cleared, the system switches over automatically to regular down-peak dispatching from the top landing.

Forgotten-man Pickup.   Included in down-peak dispatching is what is called *Forgotten-man Pickup*.   Sometimes several cars may pass a lower-landing call and cause it to go unanswered for more than a normal period.   To prevent this from becoming serious, if any call goes unanswered for more than a minute, the system books this as a forgotten call.   Then the first empty up car stops, picks up the waiting passenger and returns to the main floor landing.

Changing from down-peak dispatching to zone-return operation and back is done automatically by an electronic time totalizer.   During down-peak operation each lower-floor down call is timed from when the corridor button is pressed until the call is answered. When the total call-seconds reach a set high, indicating that more service is required in the lower zone, operation swings over to zone-return control.   When the call-seconds drops to a low set value the system changes over to time dispatching from the top terminal or near this terminal.

After the regular business-day traffic has passed, the traffic-flow switch is turned to *Night* to provide an after-hour or holiday service consistent with the needs of the building.   Night service allows the cars to remain idle at the main landing when there are no calls to answer.

When a call is registered a buzzer sounds and a highest-call return light goes on.   The attendant starts the car and it goes to the highest registered down call, answering up calls on the way up.   At the highest call the car reverses and answers down calls on the way down.

For those buildings where a full-time night attendant is not required, one elevator can be equipped for automatic nonattendant night service control.

This feature allows the elevator to be switched over to self-service operation when the last attendant goes off duty. Untrained building employes and occasional tenants can then safely operate the elevator. Adequate all-night elevator service is provided without the expense of an all-night attendant.

There are many other features of these interesting controls. But those covered give a general idea of how they watch the floor calls and take care of them automatically.

Operatorless Elevators. Haughton Elevator Company builds a system similar to the two described for automatically dispatching and operating a bank of elevators under the name of Auto-Signamatic control; K. M. White Company, markets a system called Traffic Master; Montgomery Elevator Company, Measured Demand; and Elevator Supplies Company, operatorless signal control. These systems, with a few refinements, have made possible the operation of modern elevators in a bank without an attendant, the passengers in the cars pressing the buttons for their floors.

For many years push-button controlled elevators have been developed to where they reduce the attendants' responsibilities to:

1. Pressing the car floor-call buttons at the request of the incoming passengers.

2. Initiating closing the car and hoistway doors, after which the elevator starts and automatically answers the first call in the direction in which it is going.

3. Protecting passengers against being hit by the hoistway and car doors.

Of these three only No. 3 involves the safety of the passengers. Car- and hoistway-door operators have been in use that would accelerate them to a speed of 3 fps, which constituted a hazard to passengers unless controlled by an attendant. To make these high-speed doors safe for attendantless operation their speed was reduced to about one-half or less and automatic door-reversal devices were provided. At this reduced speed, if a passenger is hit by a door the impact might cause him to be annoyed, but not hurt.

Door-reversal Devices. Four methods of reversing car and hoistway doors are available; one, a mechanical design, depends on the doors making contact with a passenger, and the other three cause reversal before the doors touch the passenger. In the first, a contact or contacts are installed inside a flexible rubber shoe running the full length of the leading edge of the car door. When a passenger entering the car presses against this shoe it deflects to close a contact that completes the opening circuit of the power operator and the door opens to its full-open position. After opening, the door immediately starts reclosing and will close unless otherwise prevented. This door-reversal device has been objected to on the grounds that it must contact a passenger to cause the doors to stop and open.

Photo-tube Protection. A second method of protecting passengers from being hit by car and hoistway doors uses light beams and photoelectric tubes. The light-sources unit is mounted on one side of the car and focused so that its beam is directed between the car and hoistway doors, across the door opening, onto a photoelectric-tube unit mounted on the opposite side of the car. These units generally move with the car and are connected into the door-opening circuit in such a way that if the light beam is interrupted while the doors are closing, the doors stop, reverse, and return to open position. This operation is similar to the familiar store or office-building photoelectric door opener and closer.

On attendant-operated cars, only one light unit is generally used. In these applications the photoelectric units protects passengers from being hit by the doors if the attendant fails to do so. On operatorless elevators two or three light beams are directed across the door opening, spaced to give full protection to children as well as adults. In these applications, when the car stops at a floor, the doors open automatically just as they do when the car is started by an attendant, but their closing is controlled by a timing device.

This device is set to permit the doors to remain open long enough for passengers to get off or on the car. At the end of the period if the light beams are not interrupted by a passenger the doors close automatically. After the doors start to close, if the light beam is interrupted, they stop, reverse, and return to open position. Should the obstruction in the light beam not clear in a set period the doors slowly start to close with the electronic protection cut out of circuit, and try to nudge the obstruction gently out of the way. At the same time a buzzer sounds and a light flashes on the supervisor's panel in the main corridor. If the obstruction does not clear the supervisor can send someone to investigate the delay.

Cold-cathode Tube Protection. A third method of protecting passengers from being hit by car and hoistway doors uses a series of cold-cathode tubes electrically connected to a metal strip antenna installed on the leading edge of the car door (Fig. 10-11). A black plastic face plate covers the entire tube assembly to protect it against being injured. These tubes create a continuous electrostatic field that extends from shoulder height to the bottom of the doors. The field curves around the edge of car and hoistway doors to protect passengers against both. This tube assembly connects into the door-opening circuit in such a way that if a passenger's body comes within 4 in. of the edge of the doors when they are closing they will stop,

FIG. 10-11. Cross section shows electronic detector installed on edge of elevator car door.

reverse, and return to open position. The doors reverse to open only when it is necessary to protect the passengers.

When the protective device causes the doors to reverse and open they return to open position, then start to close again immediately or after a short delay. If a passenger remains in the path of the doors, they will attempt to close several times, after which a warning buzzer sounds and the doors start to close slowly with the electronic device cut out. As with the photoelectric-tube device the doors try to nudge the passenger out of their way. At the some time a light flashes on the dispatcher's panel to warn him that operation of the doors is being interfered with. Pressure of the doors against the passenger is too light to cause a hazard.

Pressure Plates. Most of us have walked onto a pressure plate in front of a store door and had the door open automatically. Similar plates, known as switch mats or safety mats, are used to keep elevator car and hoistway doors from closing on passengers entering or leaving the car. These plates, made of rubber $\frac{5}{16}$-in. thick, are installed in or on the landing floor in front of and near each landing door. A passenger stepping on the plate when entering or leaving the car sets up a time-delay relay circuit that acts to hold the car and hoistway doors open. These doors remain open for a period proportional to the number of persons entering or leaving the car, thus eliminating the arbitrary door timing.

To avoid any hazard from electric shock the plate relay is powered by a 6-volt transformer. When traffic is heavy and there is a'possibility of someone standing on a mat to hold the doors open a timer may be installed. With such a timer the mat becomes inoperative in 10 to 15 sec after the car stops at the floor.

On some operatorless elevators when special protection against passengers being hit by car and hoistway doors is required, two or more of the mechanical and electronic devices described may be used.

Overload Protection. To prevent overloading operatorless elevators a weighing device may be installed in the car-sling crosshead or under the car. One of these equipments consists of a lever system held in normal load position against the compression of coil springs. If the car is overloaded, it moves down slightly, compressing the springs to actuate a contact. If this contact is in the control circuit and opens, the car cannot move until the load is reduced to or below normal. In this case the contact is normally closed.

Instead of the weighing device being used to prevent the elevator from starting under overload, a normally open contact may be used to bypass the hall calls. Under these conditions, when the elevator is overloaded the weighing device contact closes to bypass the hall calls as does the bypass switch in the car when it is closed. Then the car answers only calls made by car buttons until the load has been reduced below rated, when the car goes into normal service again.

Phantom Voice. With the marked trend toward automatic operatorless passenger elevators, what is known as a phantom voice is being used to help keep these elevators at their best performance under conditions involving the human element. The voice includes a number of tape recordings in an elevator-penthouse console. Under circumstances of interest to a car passenger an appropriate tape is set in motion past a single-channel magnetic head. This picks up the message and relays it through the elevator's control cables to a loudspeaker in the car.

If a passenger enters a waiting car and fails to register a call in a predetermined time, a tape is set in motion past the magnetic head which transmits the message "Please press your floor button." When the closing of car doors is delayed by one passenger waiting for another, a tape is set in motion which relays the message "Please release the doors." If a car becomes inoperative with the doors closed or stops between floors, the relayed message is "Be calm; a protective device is now in operation." When the elevator has been stopped by someone pressing the stop button, the message will be "Please pull out the stop button," or whatever other message may be appropriate for the circumstances for which the voice has been set. Most voice consoles have a set of six messages.

When a bank of elevators could be dispatched automatically according to traffic demands and reliable safe automatic car- and hoistway-door controls were developed, operation without an attendant in the car became a reality. This type of elevator is now used extensively in large office and other buildings. In large buildings dispatchers may still be employed, but their job is that of a supervisor, to direct passengers when needed, answer their questions, and perform other duties necessary to smooth operation of the elevator system.

CHAPTER 11

Elevator-door Operating Equipment[1]

Good Elevator Service. Good elevator service often is, fundamentally, a question of maintaining operating schedules. But operators cannot be expected to do this unless elevator hoistway doors operate easily, particularly if they are hand-operated. When hoistway doors are hard to handle, it is difficult to keep good operators, and operators that can be kept will use hard door operation as an excuse for not maintaining schedules and for neglecting their work.

Hoistway Landing-door Hangers. Hangers on the earlier type of hoistway landing doors were crude and unsuited to the service, in many cases being too light to stand up and having sleeve bearings in the sheaves. All these faults contribute to difficulties in handling and maintenance of the doors. Engineering development has produced simple, substantial ball-bearing door hangers. Figure 11-1 shows one type, on which the door is carried on a heavy steel rail rigidly mounted on the hoistway wall. Running surfaces are given a smooth finish as are those of the oversize ball-bearing sheaves of the heavy hanger. Two sheaves are welded, one near each edge of the door, to a heavy steel bar attached by cap screws to the top reinforcing bar of the door. Door-bottom clearance is adjusted by adding shims between door and hanger bar.

Well-designed modern hangers have large-diameter ball- or roller-bearing sheaves. Some bearings are lubricated by a grease gun; others are of the self-lubricating type. Hangers are built for one-way-opening single-speed and two-speed doors, and center-opening single-speed and two-speed doors, operated either manually or by power operators.

There is a tendency for doors to be thrust up when force is applied to open or close them. Accurately ground hardened-steel wheels,

[1] Assistance in the preparation of this chapter is acknowledged to the Otis Elevator Company, Elevator Supplies Company, Westinghouse Electric Elevator Company, F. S. Payne Company, and the Moline Accessories Corporation.

running on ball bearings, take this upward thrust. Various methods are used to keep these wheels in contact with the lower edge of the rail. The wheel at W in Fig. 11-1, for example, is mounted on a fitting pivoted at A to the hanger. An adjusting screw S, at the other end of the fitting, raises or lowers the sheave to give desired clearance between wheel and rail. In other types, the thrust-up roller runs on an eccentric stud which is turned for adjustment.

Two-speed and Center-parting Doors. Single doors, supported on two hangers, as shown in Fig. 11-1, are usually used on small cars and slow-speed doors. This simple arrangement requires but little operating equipment. Car and hoistway doors for medium-size elevators are frequently made in two sections, with both doors opening either from the center (center-opening doors) (Figs. 11-2 and 11-3),

Fig. 11-1.   Elevator-door hanger has ball-bearing sheaves.

with the doors opening one to each side; or from one side (two-speed doors) (Figs. 11-4 and 11-8), in which case one section slides back over the other. The center-opening type is simpler and requires only one rail, but if space is limited it will not provide any greater opening than does a single door. If door travel must be confined to car width, then the two-speed type is preferable, for this permits a door opening two-thirds of the car width. This arrangement requires two tracks, however, and operating equipment is more complicated.

Two-speed doors are never as neat in appearance as single doors or the single center-opening type, because one section must lap over the other. If unusually large door openings are required, however, a center-opening two-speed, or even a center-opening three-speed, design may be required, even though the operating mechanism is complicated. The three-speed design requires three tracks, the two inner sections passing in back of the outer one as the door is opened.

Where single-speed center-opening doors are manually operated, some means must be provided to synchronize the movement of the sections. The rack-and-pinion arrangement (Fig. 11-2) consists essentially of a pinion and two hold-down rollers, all ball-bearing mounted, centered on the rail between the doors. Each inner door hanger carries a pivoted rack engaging the pinion, thus requiring the two doors to operate synchronously.

Another arrangement for accomplishing the same result is shown in Fig. 11-3. Here the inner or front hangers are joined by a cable passing over sheaves at the outer ends of the rail. In the figure a

FIG. 11-2. Pinion and rack connection between center-opening elevator doors.

FIG. 11-3. Cable connection between two center-opening-type elevator doors.

rubber-covered cable is used to connect the hangers together and to give quiet operation. Other door hangers use silent chains.

Hangers and operating equipment are also available for two-speed center-parting arrangements (four doors, two moving in each direction), and for three-speed one-way opening doors (three doors, two passing behind the third), but these are simply more elaborate arrangements of those already described.

Manual Door Operators. Many types of horizontal manual door operators are used; that in Fig. 11-4 on a two-speed one-way opening door being typical. The operator is fastened to the high-speed door at *A* and to the low-speed door at *B*. The horizontal operating bar pivoted at *C* is attached to door closer *D* and electrical interlock *E* on the door frame. To open the door, the attendant pulls to the left on lever *F*, causing the right-hand door to slide back over the left-hand one, as in Fig. 11-5.

The inside of the door closer and interlock is shown in Fig. 11-6. Interlock contacts are in series with the car-switch or push-button circuit so that the car cannot be operated until doors and interlock are closed. This is essential for safety, because most elevator accidents occur at hoistway doors without interlocks or with interlocks made inoperative. Several states require approved interlocks on car and

Fig. 11-4. Two-speed elevator doors and manual operator in closed position.

hoistway doors. In these states the number of accidents at elevator doors is almost negligible.

The door closer (Fig. 11-6) includes a closing spring attached to the plunger and cylinder bottom. At the lower end of the cylinder is a retarding chamber into which a piston enters as the doors approach closed position. Oil in the cylinder damps closing action. When the doors open, the piston pulls up, as shown in Fig. 11-7. This puts the spring in tension and fills the dashpot or retarding chamber with oil. When the doors are released, spring tension closes them. The oil damper slows down their motion as they approach closed position to avoid noise or jar. The interlock switch in the control circuit makes contact as the doors close and breaks it as they open.

Other types of door closers compress the spring as they open (Fig. 11-7). An inner, smaller spring, not compressed until the doors are almost fully open, prevents the doors from slamming open. This small spring also assists the larger one in starting closure.

Assuming the doors to be closing (Fig. 11-7), the piston is just entering the retarding chamber, meeting the oil, and forcing it upward

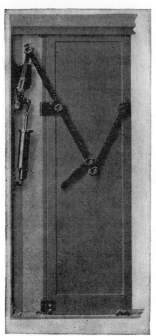  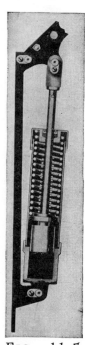

Fig. 11-5. Two-speed one-way-opening doors and their manual operator shown in the open position.

Fig. 11-6. Close-up of elevator-door closer and interlock shown in Fig. 11-4.

Fig. 11-7. Cross section through elevator-door closer.

through the annular spaces between piston and cup, thus checking door speed. This action grows more pronounced as the doors close, thus avoiding slamming. The checking effect can be adjusted by changing piston travel. Other door closers use needle valves or other adjustable devices for the same purpose.

Closing spring S is separate from oil-retarding cylinder R on the door operator in Fig. 11-8. Thus both spring and cylinder can be smaller in diameter, allowing the doors to slide past them in opening and giving maximum door opening. The oil retarder acts to resist slamming at either the fully opened or closed position. Additional safety is provided by the interlock switch, which drops open if it becomes mechanically disconnected, preventing operation of the elevator. In this it is similar to the interlock shown in Fig. 11-6.

Rack-and-pawl arrangements (Fig. 11-4) are used on elevator doors to prevent them from being forced open by someone attempting to board the car after the interlock closes. When the doors are within 4 in. of being closed, pawl *P* drops back of rack *R*.

Figure 11-9 shows a device for holding doors open. Spring-actuated composition roller *P* drops back of rack *R* on the door sill when the doors are opened and holds them until it is released by a slight pull by the attendant.

Fɪɢ. 11-8.   Door-closer spring and retarder mounted as separate units.

Hoistway- and Car-door Power-driven Operators.   High-speed elevator service demands and is dependent upon corresponding high-speed car- and hoistway-door movement. This means that these doors must be power-operated. A totally enclosed elevator car must have power-operated doors with their control tied into that of the machine. For highest efficiency in elevator operation, hoistway doors must also be power-operated and move in synchronism with the car doors. When elevator controls are automatic, at least the car doors must be power-operated and open and close automatically as the car stops or starts. For these and other reasons, rapid advances have been made during recent years in applying power operators to elevator doors.

Practically all earlier types of power operators were compressed-air engines, but modern operators are electrical types, usually motor-driven. There are two general classes: one where individual operators are applied to car and hoistway doors, the other where a single operator handles both car and hoistway doors.

Car-door Operator on Car.   Figure 11-10 shows a simple car-door operator, consisting of a motor M connected to a worm gear W which is attached to the car doors by a crank and a system of arms and levers.   Operating arm A is fulcrumed at F and connects to the doors at D_1 and D.   A connecting rod attaches the top end of arm A to the crank on the operator.   When the crank is driven in a clockwise direction, movement of arm A around its fulcrum opens the doors by passing one behind the other and moving them to the left.   Reverse movement of the operator closes the door.

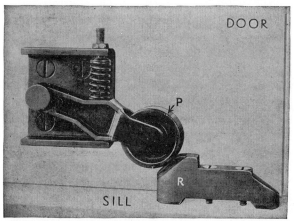

Fɪɢ. 11-9.   Device for holding elevator doors open.

Control of the door operator is tied in with that of the elevator.   As the car stops at a floor, a circuit is completed which causes the operator to open the doors.   Movement of the car-operating switch to either the up or down position completes a closing circuit for the door operator.   Returning the car-operating switch to the off position before the doors are closed will cause them to open again.   This is a safety precaution which prevents injury to passengers entering the car as the doors start to close.   This protection may be obtained automatically by the focusing of one or more light beams across the elevator entrance on photoelectric cells connected into the door-control circuit.   Should a passenger step into the path of the doors after they start to close, they will immediately stop and reopen.   The doors can be closed only when the light beam is not interrupted.   For further details on passenger protection at car and hoistway doors see page 201.

Car movement is interlocked with door position so that the car cannot start until the doors are closed. A control panel in the elevator-machine room manipulates the door operator and holds its control circuits complete until the doors are fully closed or opened. Upon approaching either open or closed position, motion of the doors is retarded by a dynamic-braking action of the operator motor, thus pre-

FIG. 11-10. Simple motor-powered operator for opening and closing car doors.

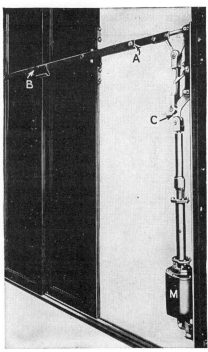

FIG. 11-11. Individual motor operator for opening and closing hoistway doors.

venting door slamming. In case of emergency the doors may be operated by hand.

Operator on Each Hoistway Door. One way of making hoistway doors power-operated is to install an operator of the type shown in Fig. 11-11 on each set of doors. This operator consists of a motor *M* connected by a driving screw to the door-operated bar. It requires space approximately equal to that taken by a standard semi-automatic door closer (Figs. 11-6 and 11-7). Control of the operator is initiated from an inductor plate mounted in the hoistway near each landing and an inductor coil carried on the car. (For a description of inductor-plate control see page 28.) When the car is stopping at a floor, the inductor coil is energized, causing the inductor switch to complete the door-operator circuit at that floor and the doors to open automatically. Moving the car switch to the starting position completes a circuit to close the doors. Closing of hoistway doors is under control of the car attendant until the car starts to leave the floor.

When the operator motor starts to open the doors (Fig. 11-11), the screw connection between motor and bell crank C is lengthened. This causes joint A in the door-operating bar to break upward and joint B to move downward, and the doors move open toward the operator. To close the doors, the motor is reversed and the operating mechanism returns to the position shown. Door operation in both directions of travel is by motor power. To prevent slamming, the door speed is checked near the ends of travel by dynamic braking of the motor. Reduced to its simplest form, this equipment is a power operator applied directly to a conventional type of door manual operator on which all oil-checking devices and closing springs have been eliminated.

Power Operators Handle Car and Hoistway Doors. Several designs of elevator-door operators use a single power unit to handle both car and hoistway doors. With these designs, hoistway- and car-door operating mechanisms are connected so that both doors move simultaneously. With the operator shown in Fig. 11-12, applied to two-speed center-opening car doors and single center-opening hoistway doors, a clutch mechanism mounted on a car door at C transmits car-door motion to the hoistway doors so that they open and close in synchronism. The operator consists of a motor M mounted directly on a worm gear. The right-hand door arm is fulcrumed to the gear case at F and the left-hand arm is pivoted at E, the whole equipment being supported from the car frame. A crank on the worm-gear shaft connects to the door-operating mechanism by a roller and link motion L. An arm extends from link L to a second link motion L' and is attached to it by a roller. The operating crank is turned in a counterclockwise direction by the motor to open the doors. When the car doors start to open, clutch C engages a roller on the hoistway-door operating bar, and the hoistway doors open simultaneously with the car doors. Full motor power is applied to start the doors; it is then quickly reduced to an amount sufficient to keep them moving at a uniform speed to the point at which slowdown begins. Two oil-check units on the operator provide cushioning action as the doors near their limit of travel in either direction.

When the car is moving up or down the hoistway, there is no contact between car- and hoistway-door mechanisms; consequently they cause no noise or wear. With other designs of car- and hoistway-door operating equipment, the power unit only opens the doors, their closure being caused by springs.

When the power operator is installed on self-leveling or automatic-landing elevators, the doors open automatically as the car

Fig. 11-12. Single power operator opens and closes car and hoistway doors.

stops at a landing, and close with the first movement of the car switch to the start position (see page 201). Opening of the doors occurs while the car is leveling, the two being so synchronized as to eliminate the passenger tripping hazard. An individual interlock is provided on hoistway and car doors to prevent the car from starting if they are not properly closed. Closing of the doors is always under control of the car attendant, and they may be stopped and reversed at any point in their travel.

Other designs of car- and hoistway-door power operators have been developed, but those described give general principles.

Freight-elevator Doors. While freight-elevator doors, as well as the doors for passenger elevators, open and close safely at the passageways between the landings and car they must be designed and built to take the heavy punishment caused by modern freight traffic. They must take not only concentrated impact loads but also the twisting torsional stresses that develop as heavily loaded trucks come across their sill. These doors are usually built of two-ply prime

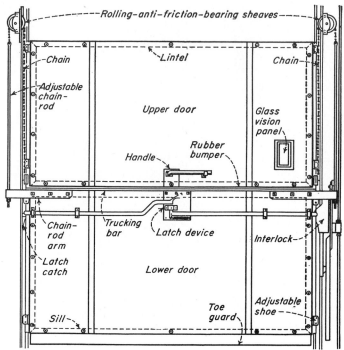

Fig. 11-13. Biparting freight-elevator doors have bottom section counterbalanced from the top section.

quality wood covered with 26-gage steel having vertical locked seams or all steel. Both designs are supported in heavy steel frames.

Whereas passenger-elevator car and hoistway doors generally move horizontally, freight-elevator doors usually open vertically to give a clear passage the width of the car for free movement of freight. When possible, these doors are usually made center parting (biparting) with the top section counterbalanced on the bottom one (Fig. 11-13). When the doors open they part at the horizontal center line, the top section moving up and the bottom down.

A rod and chain is run from each end of the trucking bar or truckable sill on the top of the bottom door section over a rolling-friction-bearing sheave and down to the bottom of the top section of the door. In this arrangement, the weight of the bottom door section is balanced by that of the top one. Since the weight of one section balances that of the other, only a comparatively small force is required to move them.

The truckable sill, of heavy steel construction, fits between the landing floor and the car platform to provide a smooth trucking surface from the landing into the car or vice versa (Fig. 11-14). This sill, when the doors are open, rests on solid adjustable stops fastened to the door guides, which in turn are bolted to the hoistway

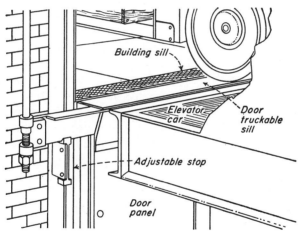

Fɪɢ. 11-14. A truckable sill between the floor landing and car platform provides a smooth trucking surface from building to car floor.

steel framing. Each sill is designed to support a trucking load equal to the elevator capacity.

A safety bumper or astragal (Fig. 11-15) on the bottom of the top door section eliminates danger of shearing fingers by overlapping steel edges when the doors close. Tension dual-side latching provides positive and parallel safety locking of upper- and lower-door panels.

Where space is not available for the bottom panel of center-parting doors to go down below the landings, telescoping or single-section doors may be installed. On telescoping doors the two sections are installed so that the bottom section can move up over the top section. Each section is guided by continuous steel members and roped together to give a 2-to-1 speed control. Both doors move simultaneously but the bottom panel travels at twice the speed of the top one. For easy movement the doors have counterweights and are supported on rolling-friction-bearing sheaves.

On small elevators, or where biparting doors cannot be installed because of space limitations below the door opening, single-section slide-up doors may be used, if sufficient headroom permits. These are also counterweighted for easy handling and always move up when opening.

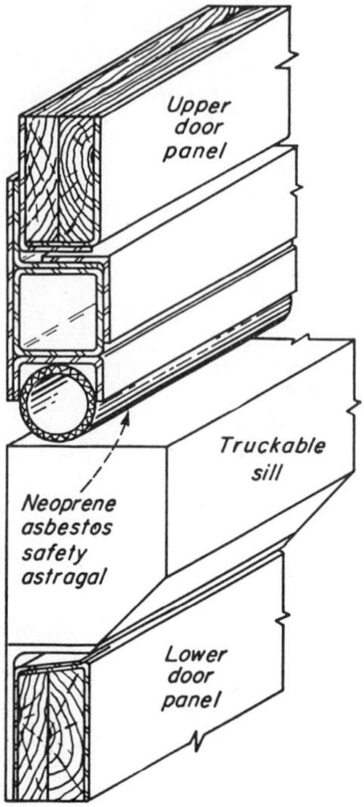

Freight-elevator car vertical-slide gates are of much lighter construction than the hoistway doors. One design is made of diamond-mesh steel welded into a rounded-edge formed-steel frame. Others have solid reinforced 22-gage panels in tubular-steel framing to give rigidity. Both are generally built in one section but, where headroom is limited, they are constructed in two sections, one moving up over the other as they open, and are counterweighted for easy operation.

Fig. 11-15. Safety bumper or astragal on bottom of top door section.

Freight-elevator Door-power Operators. Hoistway doors and car gates on freight elevators are frequently individually power-operated by motors built in as part of the sheaves over which the door or gate operating chains run. Each door may have two power units, one on each side of the door. Others have a single operator on one sheave with a cross roller-chain drive to the other (Fig. 11-16). This drive has a single heavy-duty a-c squirrel-cage torque motor M which may be stalled without being damaged if the doors become jammed partway open. Also the motor will not produce torque enough to damage the doors or operating equipment.

A control, usually located adjacent to the elevator control, selectively operates the entire line of door-power units for a single elevator. A combination zone and interlock combined with the control permits operating the door-power unit only at the floor at which the car is located, from push buttons in the car.

Car-mounted hoistway-door motor operators are available that power freight-elevator hoistway doors at each landing, and are comparable to the door operators on passenger elevators. One design has a horizontal shaft mounted above and on the car with a sprocket on each end, spaced the car width. Endless roller chains run on these sprockets and on others near the car floor and are driven by a torque motor-power unit. Mounted on one strand of each chain is a powerful electromagnet. When the car is level with the floor each magnet is opposite a bar grip on the hoistway doors (Fig. 11-17).

FIG. 11-16. This door-operator power unit is driven by heavy-duty torque motor which may be stalled without being damaged if the doors become jammed.

Energizing the magnets locks them to the bar grips so that the doors can be opened (Fig. 11-18) by the power unit that drives the sprocket shaft. When the doors are opened, the magnets release them, allowing the car to relevel if necessary at the floor without the doors being moved from full-open position.

Other designs of car-mounted door operators are similar to those for the individual door operators described, except that one operator powers each door in one hoistway, including the car gate.

On freight elevators the car gate and hoistway doors can be controlled by push buttons in the car or by the elevator control so that they open automatically as the car comes to a stop at a floor and

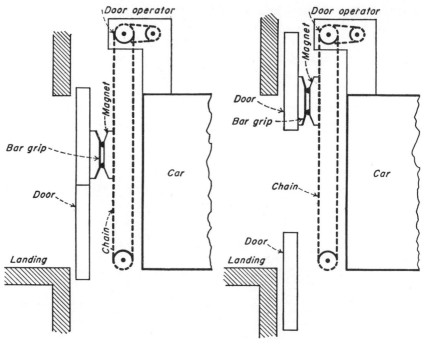

FIG. 11-17. Magnets on door-operator chains opposite bar grip on door.

FIG. 11-18. Operator has opened doors through magnet's grip on the bars.

close before the car leaves. No matter how they are operated, they have interlocks that prevent the elevator from starting unless the gate and doors are closed and locked. This provides for the same safe operation of freight elevators as of passenger installations.

Information on freight-elevator doors and gates was supplied by The Peele Company, Guilbert Inc., Security Fire Door Company, and St. Louis Fire Door Company.

CHAPTER 12

Signal Systems

Push Buttons and Bell. What constitutes an adequate elevator signal system depends on the kind of elevator service, number of cars, and their speed. Figure 12-1 shows the simplest form of a signal device, which consists of a push button at each floor and a bell located in the car or in the hoistway. When any button is pressed, a circuit is closed through the bell. For example, closing the fourth-floor button completes a circuit from + on the battery through the fourth-floor button, to the hoistway junction box, through the bell, and back to — on the battery. Such an arrangement may be used on freight elevators as a warning that someone on another floor wishes to use the car, or the bell may be used to signal the operator by ringing it a number of times corresponding to the floor number. Either method is very limited in its application and could scarcely, in the light of modern practice, be given the name of a signal system.

Push Buttons and Drop Annunciators. Some form of drop annunciator provides a means of recording the calls made by waiting passengers, and can be used for passenger or freight service where only one or two cars are in use and the service is not exacting. Figure 12-2 diagrams the connections for an annunciator and bell. When the button at any floor is pressed, the bell rings and the floor number registers on the annunciator. For example, when the second-floor button is pressed, a circuit is completed from + on the battery through the second-floor button, No. 2 annunciator coil, the bell coil, and back to the — battery terminal. This releases the drop on the annunciator, which indicates the floor and rings the bell or buzzer to call the attention of the operator. This system does not show the operator the direction in which the passengers wish to go, which limits its use to a single annunciator on installations having one elevator.

An up-and-down annunciator (Fig. 12-3) overcomes some of the foregoing difficulties. This type has two sets of indicators, one for up direction and one for down. At each floor, with the exception of the terminal landings, there are two buttons, one for up and the other for down motion. At the terminal landings there is only one button. At the top landing the button indicates down calls, and at the bottom landing the button indicates up calls.

If the fourth-floor down button (Fig. 12-3) is pressed, a circuit is completed from $+$ on the battery through button $4D$ and to terminal $4D$ in the hoistway junction box. From this box the circuit goes to the car, through annunciator coil $4D$ and the bell coil, returning to the $-$ terminal of the hoistway junction box and the battery. Energizing this circuit releases the annunciator drop and rings the bell, the latter attracting the operator's attention to the call.

A passenger pressing the second-floor up button would make a circuit from the $+$ battery terminal through the $2U$ button to the $2U$ hoistway-junction-box terminal. From here the circuit continues through the $2U$ annunciator coil and the bell coil, returning to the $-$ battery terminal. This operation releases the $2U$ indicator on the annunciator and rings the bell, again calling the operator's attention. From the foregoing, it is seen that this system not only calls the operator's attention to waiting passengers, but also indicates the direction in which they wish to be taken.

On account of the greater information given by this system, it is sometimes used on two passenger elevators serving buildings where the service is not too exacting. When a double-drop annunciator is used on each of two elevators, the same push-button arrangement on the floors is used as shown in Fig. 12-3, and the annunciator in one car is connected as shown in the figure. The annunciator in the other car connects to a junction box in the hoistway, also shown in Fig. 12-3, and the two junction boxes are connected in parallel. For instance, terminal $5D$ of one junction box connects to $5D$ in the other, $4U$ to $4U$, $4D$ to $4D$, and so on. With this connection, when any one of the floor buttons is pressed, the call will be registered on both annunciators.

Sometimes the annunciator on each car is connected to the same junction box in one of the hoistways. The objection to this arrangement is that trouble on one car may interfere with the operation of the other car. If the junction box is in No. 1 car hoistway and No. 2 car signal system is in trouble, testing from the junction box to No. 2 annunciator will require stopping car No. 1 to make the test. Also, it will be necessary to shut down car No. 1 to renew the traveling cable of No. 2 annunciator. For these reasons, it is best to provide

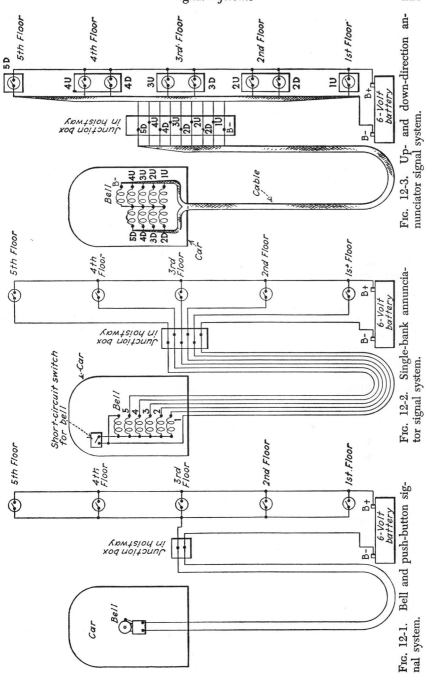

FIG. 12-3. Up- and down-direction annunciator signal system.

FIG. 12-2. Single-bank annunciator signal system.

FIG. 12-1. Bell and push-button signal system.

a junction box in each hoistway and connect these in parallel, with a short length of cable.

Objections to Annunciator Systems. One objection to the annunciator system is that there is no means of indicating to the operator of one car when the other car has answered a call. The annunciator does not automatically reset when a call is answered. Since they should not be reset until the operator has answered all calls in one direction, there is no means of recording a call after the car has stopped at a floor. For instance, assume that the car has a call to stop at the third floor up and after answering this call continues in the up direction; if after the car has left the third floor and before the annunciator is reset at the top floor a passenger on the third floor presses the up button, this call will not be recorded on the annunciator. The drop for the third floor is in the call position from the previous call. When the operator, on reaching the terminal landing, resets the drops, no indication of the third-floor call is left. This difficulty is inherent in all drop annunciator systems and for this reason they are usually limited to single-elevator installations, although in some old apartment houses annunciators of this type are used on two-elevator systems, and on elevators serving fifteen or more floors.

Bells Used on Annunciators. The type of bell used on the annunciator may vary with conditions and the ideas of the person making the installation. Although a continuously ringing bell has the advantage of being certain to attract the operator's attention to the call, it may be very annoying to the car passengers if someone at a land ing persists in pressing the button until the arrival of the car. For this reason, a short-circuiting switch (Fig. 12-2) is sometimes connected to the terminals of the bell, so that if the bell becomes annoying it can be short-circuited and only the annunciator drops will operate.

A buzzer is equal to a bell in attracting the operator's attention, is not nearly so annoying to passengers, and is used extensively. The single-stroke bell is probably equal to any other noise device for attracting the operator's attention and is used widely.

Power for Signal Systems. Where no other source of power is available, dry cells are used extensively for operating the annunciator signals. These have the disadvantage of failing at inopportune times if not given attention and renewed before completely run down. They possess the advantages of being readily available and easily and cheaply replaced. Where a storage battery is available for operating other low-voltage devices, the elevator signal system may be connected to the number of cells in this battery that will give the desired voltage. Although motor-generator sets are used extensively to operate the more elaborate signal systems for elevators, they are rarely used on such systems as previously described. Where alternating current is available, bell-ringing transformers are frequently used. On account of the inductance of the circuits, an a-c voltage about 100

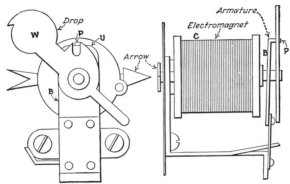

Fɪɢ. 12-4. Diagram of annunciator drop, indicating arrow in off position.

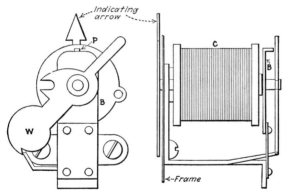

Fɪɢ. 12-5. Diagram of annunciator drop, arrow in indicating position.

per cent higher than the d-c voltage should be used. That is, if 6-volt direct current is used, about 12- or 15-volt alternating current will be required. On some systems, 8-volt direct current and 24-volt alternating current are recommended.

Types of Annunciator Drops. A great variety of drops have been developed for use in annunciators. That in Figs. 12-4 and 12-5 shows the general principle. An arrow fastens to one end of a shaft that extends through the coil, and on the other is a weight W, arranged to turn the shaft about 90° when free. In off position a pawl *P* holds the weight arm as shown in Fig. 12-4. When coil *C* is energized by the closing of a push button, it attracts armature *B* and releases weight arm *W*, which drops to the position shown in Fig. 12-5. In so doing, the weight arm turns the arrow to a vertical position and gives an indication of the call. When the button is released, it de-energizes coil *C* so that armature *B* returns to normal position.

FIG. 12-6. Gravity-type drop which is reset mechanically.

FIG. 12-7. Electrically reset drop.

FIG. 12-8. Up-and-down type of annunciator using drops shown in Fig. 12-7, but mechanically reset.

Drops of this type are reset mechanically by the operator pushing up on a rod extending below the lower end or through the front of the annunciator case.

Where arrows give the indications, the only change on the annunciator is the position of the arrows, which probably does not possess as good qualities for attracting attention and being interpreted correctly as does a change of color or the flashing of a number. To make the calls on annunciators more readily seen and correctly interpreted, a number of methods have been devised. One of these, shown in Fig. 12-6, is so designed that when coil *C* is energized, the white target drops down in front of a transparent opening. With this arrangement, the only indicators exposed are those representing calls, and as these are shown by figures against a white target they are easily read.

Another method exposes a white target in front of a transparent opening, with the floor numbers painted in white above the openings for the targets. Figure 12-8 shows a Holtzer-Cabot Electric Company annunciator of this type, and a drop with a reset coil is shown

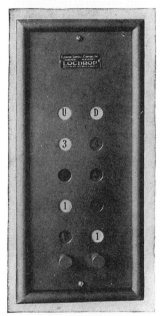

FIG. 12-9. Annunciator with drops exposed back of the floor-call numbers.

FIG. 12-10. Floor-call annunciator shown in Fig. 12-9, with the cover removed.

in Fig. 12-7. When coil *B* is energized by the pressing of a hall button, the lower end of armature *A* is attracted and the target drops to the right and is exposed under the floor number. When the reset button is pressed in the car, coil *C* is energized and the target is moved to the position in the figure.

Another system used in elevator annunciators is that of Elevator Supplies Company, shown in Figs. 12-9 and 12-10. Floor numbers are painted in black on transparent glass; then, when the white targets are exposed behind the glass, the numbers show up clearly, as indicated in Fig. 12-9. The *U* and *D*, indicating up and down direction, respectively, have a white background painted on the glass behind the letters. In Fig. 12-8 the annunciator is arranged for mechanical reset by the pushing up on a knob extending below the lower end of the case, while in Fig. 12-9 the reset knobs are horizontal and are located just below the indicators. Figure 12-10 shows the same annunciator as Fig. 12-9, with the cover removed. Targets when not exposed are located on the left-hand side of the coils. When the coils are energized, the targets are rotated through 90° into the position shown in white on three of the coils. Figure 12-11 shows a push button used with the annunciator shown in Fig. 12-9, and a push button for a single-bank annunciator is shown in Fig. 12-12.

Fig. 12-11.   Fig. 12-12.

Fig. 12-11.   Up and down annunciator floor push button.
Fig. 12-12.   Single-bank annunciator floor push button.

Electric Reset Annunciators.   Although mechanical resets are used on elevator annunciators there are a number of objections to them. Electric reset annunciators are used extensively and eliminate some of these objections.   On these types two coils are generally used on each drop, one to give the indication and the other to reset the drop. Reset coils on a single-bank annunciator are connected in parallel to the push button shown in Fig. 12-13, located in a convenient position for the operator.   When the operator presses the button, all drops are reset.   Up-and-down type annunciators have two reset buttons, one for the up bank of drops and one for the down bank.   Such annunciators can be located in any position most easily observed by the operator; they are not subjected to the abuse of the mechanical reset type and are quieter in operation.

Electric reset requires that both battery wires be brought to the car, but most annunciator cables are provided with an extra wire that can be used for this purpose. On some annunciators the bell or buzzer is not connected in series with the annunciator coils (Figs. 12-2 and 12-3), but is connected to the contacts of a relay (Fig. 12-14). This connection also requires that both battery wires be brought to the car, where they can be used for the electrical reset, too. Putting in a relay (Fig. 12-14) takes the buzzer out of the drop-coil circuits, and the only make and break in these circuits is the push button.

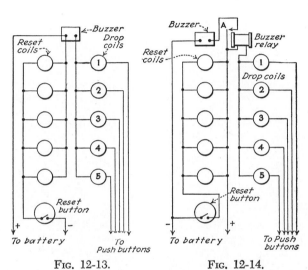

FIG. 12-13.   FIG. 12-14.

FIG. 12-13.   Diagram of single-bank annunciator with electrical reset coils.
FIG. 12-14.   Diagram of single-bank floor-call annunciator with electrical reset coils, a buzzer relay, and reset button.

In Fig. 12-14, if a hall button is pressed, it makes a circuit from the + battery terminal through the relay coil, drop coil, and push button that has been closed, to — on the battery. The relay contact A closes and completes the buzzer circuit from the + to the — side of the battery. It does not make any difference whether contact A closes or not; the drop coils will be energized whenever their circuits are completed.

Individual-drop reset annunciators have been used to a limited extent. These require a push button in the car for each drop in the annunciator. Where it is desirable to have individual reset, automatic-reset machines are generally installed and light annunciators of some type are used.

Lamp-annunciator Signal Systems. Although drop-type annunciators are used in elevator signal systems, they do not possess all the characteristics to make them well adapted to modern systems. Lamp indicators in various forms have come into use, as they occupy less space for the same number of floors than the drop type and, when the lights flash on, attract the eye to a greater degree than any of the drop methods.

With a lamp annunciator some method must be provided to hold the lamp circuits closed, after the hall buttons have been pushed, until

Fig. 12-15. Front view of a relay-type push button.

Fig. 12-16. Rear of relay-type push button shown in Fig. 12-15.

the call has been answered and the signal reset. This can be done in two ways. In one, a relay-type push button is used which, when pushed, remains closed and completes the indicator-lamp circuit until released by the reset coils located in the push buttons. In the other, the push button, when closed, energizes a relay, the contacts of which light the indicator lamp. With this system the relays are arranged in a central group, which is generally located in the penthouse, and there must be a relay for each up or down signal; that is, if there are ten floors there will be eighteen relays. When the indicator-lamp circuit is closed through a relay's contacts, it remains closed until the relay is reset by the reset device.

Relay-type Push-button Systems without Automatic Reset. Figure 12-17 is a wiring diagram of a Holtzer-Cabot Electric Company's system using the former principle. Reset coils in this system are mounted with the push buttons, as at *C* in Fig. 12-16. One of the two reset coils is for the up-direction signal and the other for the down. On each push button there are three contacts, one for the buzzer circuit, one for the signal-lamp circuit, and one for the reset coil. The operation of the buttons will be understood from *A* and *B* at the bottom of Fig. 12-17. When a button is pressed, a pawl *P* (Fig. 12-16) drops in back of a collar on the button and prevents the drop and reset contacts from opening, as indicated by the position of up button *B*. This keeps the lamp contact closed until reset coil *C* is energized and pulls pawl *P* clear of the button, which allows it to return to normal position, as in *A*.

The buzzer circuit is closed only as long as the push button is held all the way in. As soon as the button is released, it returns far enough toward its normal position to open the buzzer contacts without opening the light or reset contacts. This is the position of the up button in *B*. Every time the button is pressed to the full-on position, the buzzer will ring.

Assume that, in Fig. 12-17, the third-floor down button is pressed to the full-on position. Then contacts *A*, *B*, and *C* will be closed. These close a circuit from + on the power supply to the connection box in the hoistway and to the third-floor down push button. From the push button one circuit is through contact *A*, into wire *D* and through lamp 3 of the down bank in the annunciator, to — on the power source. Making this circuit alive lights lamp 3, which shows the location of the call. Contact *B* is connected to reset coil *E*, the circuit for which is open at reset button *G* on the annunciator in the car.

As long as contact *C* of the landing button is closed, a circuit is made through the buzzer to — on the power supply. The sound of the buzzer attracts the operator's attention to the call. As previously explained, the buzzer contact remains closed only as long as the push button is held in the full-on position. As soon as the person making the call releases the button, contact *C* opens and the buzzer stops sounding. Closing push button *G* on the annunciator completes a circuit for reset coil *E*, from *B*, through this coil, push button *G*, and to the — line. Energizing coil *E* causes it to pull the pawl away from the push button and allows the button to return to the normal open position as shown in the figure, and the third-floor down light on the annunciator goes dark.

It can be seen from Fig. 12-17 that all the down reset coils are connected to the same wire, so that when reset button *G* closes, all the down reset coils are energized and any of the down push buttons that are closed will be released. Operations similar to those which have been explained for the down buttons apply to the up, with the up lamps lighting on the annunciator instead of the down lamps.

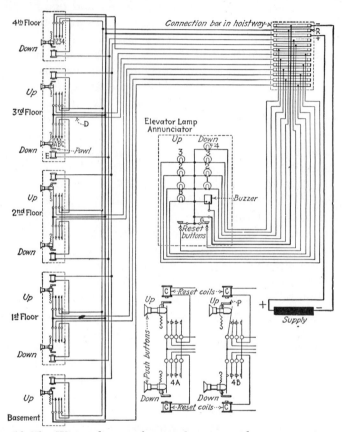

FIG. 12-17.   Wiring diagram for signal system without automatic reset.

Relay-type Push-button Systems with Automatic Reset. If this equipment is applied to two elevators, each machine must be wired to its junction box in the hoistway in a way similar to that shown in Fig. 12-17. Then it is only a matter of connecting the two boxes in parallel; that is, the — terminal of one box is connected to the — of the other, *R* of one to *R* of the other, etc. Only one machine junction box is wired to the floor push buttons, as only one set of these is used for a bank of elevators.

Although this system is simple in its operation, it lacks the individual reset feature which clears the call on the annunciator when it has been answered. This feature can be obtained by a reset equipment (Fig. 12-18) driven from the elevator machine. A wiring diagram similar to that shown in Fig. 12-17, but with automatic reset, is shown in Fig. 12-19. On this diagram it can be seen that if any of the floor buttons is pressed, it will close circuits similar to those

Fig. 12-18. Machine automatically resets floor-call annunciator drops.

shown in Fig. 12-17, light the annunciator lamps, and sound the buzzer. For example, if the down button on the second floor is pressed, circuits will be energized as explained by Fig. 12-17.

The reset machine (Fig. 12-18) is a dial switch geared to the elevator machine by a reduction gearing and chain drive that allows switch arm *A* to make, on high-rise machines, approximately one revolution when the car travels the length of the hoistway. On low-rise machines the contact arm makes less than one-half revolution, as it does on the machine in Fig. 12-18. There are two sets of contacts, *B* and *C*, on this switch, one for the up reset and one for the down. The arm carries two contacts, *D* and *D*, only one of which rcsts on the stationary contacts at a time. The two contacts on the arm are pivoted in such a way that when the arm moves in up direction, the up contact is in service. When the car's direction is down, the other contact completes the down reset circuits, and the up contacts do not close. This is necessary, since with the car in the up motion only up calls are answered and only these should be reset when answered. Likewise, on down motion only down calls are

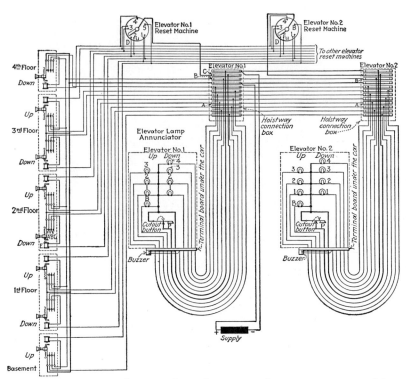

Fig. 12-19. Wiring diagram of signal system with automatic-reset machine.

answered and only these should be removed from the annunciator when answered. On traction machines, slipping of the ropes on the driving sheave tends to throw the reset machine out of the correct relation to the landings. To overcome this difficulty, if the arm is out of adjustment at the limit of travel it strikes a stop, and a clutch in the drives slips to bring the contact arm into the correct position.

Assume that the second-floor down button is pressed (Fig. 12-19). This will close push-button contacts A and B until the call is reset, and contact C as long as the person making the call holds the button in the full-on position, as previously explained. Contact *C*, when closed, completes a circuit from the + power source, to the connection board in the hoistway, through contacts *A*, *B*, and *C* of the second-floor down button, and back to terminal *B* in the hoistway connection boxes. From these boxes circuits go to each car, through the buzzer, and back to the connection boxes in the hoistway and to the — on the power source. In this particular installation the annunciator is wired to a terminal board under the car, where the buzzer is also located.

The second-floor down-landing lamp circuit is through contact *A* on the push button to *A* terminal in the hoistway connection boxes. From *A* in each hoistway connection box there is a circuit to each car, through down lamp No. 2, and back to the — on the power source. This operation lights the second-floor down lamp on each car, indicating to the operators the floor at which the call has been made.

If No. 1 car is going down, when it reaches the second floor the operator should stop and answer the call. With the car in the down motion the contact arm of No. 1 reset machine is moved until it rests on contact *2D* at about the time the car reaches the second floor. With the arm of No. 1 elevator reset machine on contact *2D*, a circuit is completed through contact *B* on the second-floor down button and reset coil *E*, to contact *2D* of the reset machines. If the arm of machine No. 1 is resting on this contact, the circuit is completed through the arm to terminal *C* in the hoistway connection box. From this terminal there is a circuit to push button *P* in No. 1 car back to the connection boxes in the hoistway and to the — on the power source. Completing this circuit energizes the second-floor down reset coil *E* and causes it to release its push button, and down No. 2 lamp on each annunciator goes dark, indicating that the call has been answered. What has been said regarding the second-floor down button applies to any of the other call buttons.

Cutout button *P* can be used by the operator to prevent the reset machine from removing the signal from the system. For example, assume that No. 1 car is coming down and that when it leaves the third floor it has a full load of passengers. When the car passes the second floor, the reset machine will clear the down signal for this floor. To prevent this, the operator holds button *P* open, which breaks the reset circuit and prevents the second-floor down signal from being cleared. The signal remaining on No. 2 annunciator shows the other operators that the second-floor down call has not been answered.

Signal Systems Using Conventional Push Buttons. On the two signal systems described, the relays are built into the landing push buttons. Several systems have been developed that use conventional push buttons and separate relays.

The reset machine for such a signal system consists of two horizontal rows of contacts, as shown by A and B in Fig. 12-20; one is for the up and the other for the down signals. Brushes C are moved over each set of contacts. When the car is in down motion the brushes are in contact with the down contacts only. On up motion the brushes shift from the down to the up contacts. By this arrangement signals are reset only in the direction in which the car is going.

The reset machine is driven by a sprocket chain from the end of the drum or tractor-sheave shaft. Brushes are carried on a screw

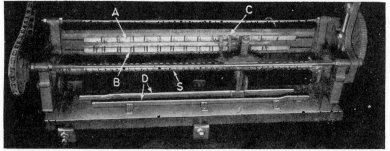

Fig. 12-20. Horizontal machine for automatically resetting the signals after the elevator has answered the floor calls.

shaft S. On this shaft the sprocket has a clutch that slips to compensate for slippage of the ropes on traction-elevator machines. If the ropes slip and throw the reset machine out of adjustment, the screw shaft runs against a stop before the car reaches a terminal landing and is prevented from turning. For remaining car travel to the terminal, the screw-shaft clutch slips. In this operation, brushes are brought into correct adjustment. The brush carriage rides directly on the driving screw and is prevented from turning with the screw by stops D. These stops are set so that when the direction of the screw is reversed the brush carriage is moved far enough in the screw's direction of rotation to transfer the brushes from one row of contacts to the other.

Only one bank of relays is required regardless of the number of cars, but a reset machine is needed for each car. When a floor call button is pressed it registers a signal on the annunciator of all cars. The reset machine on the first car answering the call removes the signal from the annunciator in all the cars. Thus the operators are kept informed of the calls to be answered.

Signal systems of this kind are used for groups of one or more elevators in buildings, such as hotels or office structures, where the service may be intermittent and the elevators do not always make complete trips. Where the elevators are intended to make complete trips, an operator's signal-light system is frequently used. Details of other types of signal systems are given in Chaps. 9 and 10 on signal control with microleveling.

Elevator-car Schedules. A building may have a sufficient number of elevators, but if they are not used efficiently, the service will be unsatisfactory. An adequate signal system is essential to the operation of modern elevators. Operators must be informed as to where the waiting passengers are and the direction in which they wish to go. It is also necessary for waiting passengers to know when cars are approaching and the direction in which they are going.

It is of equal importance to operate the cars on a proper schedule. Unless this is maintained at all positions in the hoistway, the service will be erratic—too many cars may be passing a given point in the hoistway, while at other points the service may be inadequate. Service may be irregular even when the cars are controlled at the ground floor by a starter.

Automatic Car-dispatching Systems. In recent years considerable attention has been given to relieving the starter of dispatching the cars by making operation automatic and providing means to show the operators if they are running on schedule. Such a system of automatic elevator control has been developed by the Elevator Supplies Company. In each car there is a three-light unit (Fig. 12-21) containing a green, a red, and a white light. The green an the red lights are for dispatching the cars at the terminals, and the white lights show the operators if the schedule is being maintained. The machine for operating the system is shown in Fig. 12-22; it consists of three motor-driven arms A for flashing the lights. In the top of the cabinet are duplicates of the green and the red lights in the cars. Below these lamps is a set of starter's call-back buttons, which are connected to the green light in each car. The dispatching cabinet is near the starter's station on the ground floor.

The dispatching machine is designed for scheduling a group of three to five elevators. That in Fig. 12-22 is designed for scheduling four elevators. For a group of four elevators, with contacts C placed in the end of the arms, they will make contact with buttons 1, 2, 3, and 4 and give signals to all four elevators. If the arms are properly spaccd the cars will be dispatched on equal time for either direction. For example, if the round-trip time is 180 sec the cars will be signaled to start at 45-sec intervals.   When it is time for elevator 1 to start a green light will flash in that car.   When it is time for the operator to leave the top floor a red signal will flash in his car.   This will be 90 sec after the car leaves the ground floor.   When the car

Fig. 12-21.   Car operator's dispatching signals.

Fig.   12-22.   Machine that automatically schedules a bank of elevators.

should be halfway down the hoistway the white light will flash.   If the car is above the halfway point, the operator knows that he is behind schedule.   On the other hand, if the car is below the halfway point the signal shows that the schedule has been exceeded. Forty-five seconds after No. 1 car leaves the ground floor the operator in No. 2 car will receive a green signal to leave.   Similarly, 45 sec after No. 2 car starts, No. 3 will get a signal to go.   This system of equal up and down time is the one that is normally used between rush periods. With this setup the arms that flash the green, the red, and the white signals make one revolution in 180 sec.

In rush periods the up and the down traffic is out of balance. During the morning-in and noon-in periods traffic is heavy in the up direction, while during the noon-out and evening-out periods the traffic is largely in the down direction. To take care of these conditions the red and the white lights can be adjusted to allow a longer running time in the direction of heavy traffic. For instance, if the round-trip time is 180 sec, the arms on the automatic dispatcher may be so placed that the time in the direction of heavy traffic will be 100 sec, and that in the opposite direction 80 sec.

Instead of the white lights being used to give an intermediate signal to the operator, they may be employed to give a preliminary starting signal at the lower terminal landing. When used for this purpose, the white light flashes about 4 sec before the green light. Generally a buzzer is arranged to ring when the green starting light flashes. It has been found that the chance of an operator missing a signal at the ground floor is many times that of his missing it at the other floors, on account of the heavier traffic at the former. Where the cars run to a basement floor the white light may be used to signal the leaving time from this floor and the green light used for the usual purpose at the ground floor.

The schedule may be speeded up or slowed down by adjusting the position of knob K in schedule selector S, as shown in Fig. 12-22. This adjustment is simple and can be easily made. If a car is taken out of service, the schedule can be quickly adjusted for the lesser number of cars by rearranging the contact keys C in the numbered sockets. A chart on the inside of the machine's enclosure door gives the socket numbers for dispatching the different numbers of cars that may be in service.

The green and red lights at the top of the dispatching cabinet, as previously mentioned, correspond to those in the different cars and flash in synchronism with them. These show the starter if the car operators are getting the signals to start at the terminal landings. The call-back buttons are connected to the green lights in the cars. When a button is pressed, it causes the green light in the car, corresponding to the button, to flash, and at the same time the operator receives an audible signal. Code calls may be worked out between the dispatcher and operators to suit the particular conditions under which the cars operate.

The foregoing gives a general idea of signal and dispatching systems as found in older and medium-size buildings. Modern signal systems are part of the elevator control discussed in Chaps. 9 and 10. In these systems, when a hall or car button is pressed the call is registered in the control system. The car then stops at the floor automatically, whether operated by an attendant or not, and the doors open.

Passenger Information. For elevators to operate efficiently, the signal system must give the prospective passengers the necessary information for them to know that their calls are being answered. This intelligence begins with the hall buttons, one type of which is of a mushroom design made of transparent plastic with a small incandescent lamp behind it. When the button is pressed and the call

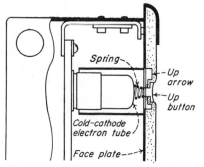
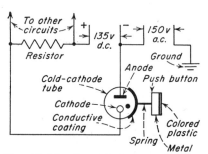

Fig. 12-23. Simplified cross section of an electronic floor-call-signal touch button.

Fig. 12-24. Simplified wiring diagram of an electronic floor-call touch-button circuit.

recorded in the control system, the lamp lights and illuminates the button until the call is answered.

Other designs of hall buttons have transparent plastic direction arrows above or below them with a lamp behind. When the button is pressed, its lamp lights to illuminate the corresponding arrow. Either one of these methods is an effective way of showing the waiting passengers that they are getting immediate attention.

Electronic Push Buttons. An electronic push button (touch button) by Otis is one of the latest developments in its field. A cross section of one is shown in Fig. 12-23. It consists of a cold-cathode electron tube, located behind a translucent direction arrow around a stationary opaque button held in a face plate. The face of the button is covered with insulating material. Its back is a conductor that makes contact through a spring with a conductive coating on the tube's top. Thus the tube forms one plate and the dielectric of a condenser. Figure 12-24 shows a simplified wiring diagram of an electronic touch-button circuit. There is 135-volt d-c potential across the tube's anode and cathode. This potential is applied from the + terminal, through the resistor, to the cathode and anode, to the — terminal, but it is not high enough to start current flowing through the tube. There is also 150-volt a-c potential between the anode and ground. For one polarity of the alternating current the voltage is applied from the right-hand terminal to ground, through the person touching the button, and the tube, to the left-hand terminal of the voltage source.

When a prospective passenger touches the button, electrostatic field distribution inside the tube changes enough for the tube to ignite and start conducting. After the prospective passenger takes his finger off the button, the 135-volt direct current maintains current flow through the tube once it is started. Current flow through the resistor causes a difference of potential across its terminals that registers the call in the elevator-control system.

When the tube ignites, it lights and illuminates the translucent arrow around the button, indicating that the call has registered in the control system. When the call is answered, a voltage pulse applied to the circuit momentarily reduces the potential across the tube to the point at which it ceases to conduct and the registered call is cancelled.

Car-position Indicators. After the passengers in the halls know that their calls will be answered, the next move is to indicate when and by which elevator it will be done. On the older elevators this was and is still done by an arrow rotating across the face of a dial, on which each floor is numbered. The arrow is usually driven mechanically from the elevator by a metallic cable or tape system, which is one objection to it. It is also limited as to the number of floors it can serve, and the arrow is generally not easily visible when viewed from a distance or at an angle.

As a result electrical hall car-position indicators have come into wide use. A common design is made in two sections, each faced with a glass lens, with a red lamp behind one and a green lamp behind the other. One of these is placed above each corridor or hall door, with the green part uppermost and the red down, connected into the signal system so that it lights when the car approaches the floor in answer to a signal. The green lamp lights when the car is to stop in answer to an up-direction signal and the red lights for a down stop signal. When the car leaves the floor the light goes out.

A single-stroke bell is sometimes combined with these lanterns to sound once when the lamp lights to call waiting passengers' attention to the approaching car. Lenses project from the wall to provide both front and side illumination, making them suitable for use with a row of elevators.

When it is deemed advisable to give waiting hall passengers complete information on the position and direction of travel of each car in a bank, a car-position indicator by Westinghouse (Fig. 12-25), is installed over each hoistway door. These indicators have lamps behind glass floor numerals arranged in a horizontal row. The lamps are illuminated individually as the car travels up or down the hoistway by contacts on the selector in the control room. Illuminated arrows, green for up and red for down, on the car position indicator show direction of car travel.

FIG. 12-25. Landing car-position indicator shows what floor the car is near and direction of travel.

A position indicator like that shown in Fig. 12-25, minus the direction arrows, may be installed in the car above the entrance door. It has amber numerals that light individually to indicate the car's position in the hoistway. This fixture thus serves to keep passengers informed of the car's progress and prompts them to be ready to leave when the car arrives at their respective landings.

Another development by Westinghouse combines an up and down hall push button with a miniature car-position indicator. Figure 12-26 shows a single-car station and Fig. 12-27 a two-car combination, but the stations are available for three cars. Floor indicators are small translucent cylinders with the floor numbers on them, behind openings in the push-button face plate. A lamp inside the cylinders illuminates the floor numerals so that they can be readily seen. The indicator cylinders are operated from the floor selector in the control room. As the car approaches a floor, its number flashes on the position indicator.

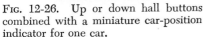

FIG. 12-26. Up or down hall buttons combined with a miniature car-position indicator for one car.

FIG. 12-27. Up or down hall buttons combined with a miniature car-position indicator for two cars.

Direction of car travel can be determined by the order in which floor numbers show up on the indicator. When the car goes up, its floor numbers are seen in ascending order, and on down travel, floor numbers flash on in descending order. A mechanical type of car-position indicator in the push buttons, by Haughton, is also available.

CHAPTER 13

Locating Faults in the Mechanical Equipment

Noisy Operation. Mechanical troubles on electric elevators are not so numerous as are electrical difficulties and in general they are not of such a nature as to cause an immediate shutdown of the machine. However, these defects are frequently of great importance in safe operation. Therefore, the mechanical parts should be given as careful attention as the electric parts.

Mechanical troubles that can develop on an electric elevator depend on the type of machine. The geared drum machine (Fig. 13-1) and the geared traction machine (Fig. 13-2) both have the same kind of motor and gears, and what is said regarding one applies to the other. Noisy motor operation was a common defect with the older designs, due to the rotating elements being out of balance or bearings being allowed to wear out so as to let armature or rotor rub on the pole piece or stator core. The latter was more likely to occur with induction motors than with d-c machines, on account of close clearance between rotor and stator of a-c machines.

Another cause of noisy operation is the armature or rotor working loose on the shaft. In elevator service the motor's rotating element is subjected to reverse stresses, due to starting in one direction and then in the other, and braking during stopping. Any slight movement of armature or rotor on its shaft will soon wear to allow considerable motion. Any movement between rotor and shaft should be corrected as soon as possible by shimming up under the key or by fitting a new key. If the repair is not made at an early stage of the trouble, both key and key seats will wear, making it necessary to recut the key seats and use a larger key. If the rotor does not have a large shaft and a substantial key, it may be difficult to prevent it from working loose. Most motor rotors are now pressed on and keyed to their shafts.

The motor out of line with the worm shaft is a common cause of noisy operation. This is dangerous if the motor is much out of alignment, since the bending stresses in the rotor shaft at the brake-wheel coupling may break the shaft. If this should happen with a heavy load in the car when the car is in down motion, the motor would be acting as a brake to keep the load under control. Should the rotor shaft break, the load would be released, and since the mechanical brake would not be applied until the operator centers the car switch, there is a possibility that the car may reach a high speed or even get out of control or car safeties set before the operator acts.

Fig. 13-1.   Drum-type elevator machine with controller mounted on motor.

Brake Wheel and Brake.   Brake and brake wheel should be given careful attention, for it is this part of the machine that stops the car, and if the brake is not in proper condition, not only is elevator service impaired, but a hazard to life and property exists.

Ascertain if brake wheel and coupling are properly secured to the motor and worm shaft. Two common troubles with a brake are that it does not stop the car quickly enough or that it stops it too quickly with an unpleasant jar to the passengers. Failure of the brake to stop the car within proper limits may be caused by a worn or oil-soaked and dirty brake lining. Putting a new lining in the brake shoes should not be attempted except by an experienced mechanic. Unless the lining makes a good fit with the shoes, it will have an uneven surface and will touch the brake wheel at only a few points, causing poor braking action.

When brake linings are in good condition and the brake does not stop the car properly, the cause is in the mechanical adjustments or in the dynamic-braking circuit in the controller. Adjust the brake so that when it is released, the shoes just clear the wheel. Then adjust tension springs or weight position to give the desired stop.

Where dashpots are used on the brake to control the rate of application, keep them clean and in good condition or their action will be erratic, which will cause differences in brake applications. If the dynamic-brake circuit does not close during stopping, with the same

Fig. 13-2. Geared-type traction-elevator machine.

mechanical brake adjustment, the car will slide. Adjusting the mechanical brake to stop the car within reasonable limits under this condition will cause an unpleasant jar. A short circuit in the dynamic resistance will stop the car too quickly, and this will generally be accompanied by sparking at the motor's brushes. Therefore, when trouble is experienced with a brake, consider the electric circuits as well as mechanical features.

On some types of brakes it is possible to apply one shoe with greater force to the brake wheel than the other. A brake adjusted this way stops the car suddenly in one direction and allows it to slide in the other.

On a-c equipment with two-speed motors, when stopping the high-speed motor is cut out and the slow-speed motor is cut into service. The latter acts as a brake to slow the machine down to correspond to the motor's normal slow speed. If the control equipment gets out of adjustment, the slow-speed motor may not be cut into service for slowing down. Then the machine must be stopped by the mechanical brake, and it will be difficult to make good stops.

Thrust Bearings. On worm-gear drum or traction machines, thrust bearings may cause trouble, particularly in the older types. These bearings may be of the disk type (Fig. 13-3) made up of bronze and steel disks, or of the ball type (Fig. 16-2). If the disk thrust is properly designed and installed and kept well lubricated, its wear should be relatively small. Whatever wear there is must be taken up by the adjusting screw. If the wear is all on the back thrust bearing, then each time adjustments are made the worm and motor shaft are brought back to their original position. When the wear is on the front thrust bearing, then the worm shaft and motor shaft are

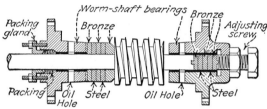

Fɪɢ. 13-3.   Worm shaft, showing bronze- and steel-disk thrust bearings in place.

being pushed toward the motor's outboard bearing. In this case care must be exercised to see that the motor shaft is not pushed so far endwise as to transfer the thrust from the thrust's to the motor's bearing. If the outboard bearing of the motor is allowed to take the thrust, it may heat, and in severe cases the thrust may be sufficient to break the bearing bracket.

On some of the older type of machines one thrust bearing was put on the motor's outboard bearing, but on these machines the motor was designed for such service. This arrangement did away with the thrust bearing on the front end of the worm shaft, which cannot be reached without taking the worm shaft out, a job that requires considerable time. With ball or roller thrust bearings, one of the chief difficulties was chipping of the balls or rollers. These troubles have been largely overcome in modern designs, in which both bearings are placed on the outboard end of the worm shaft. In addition to the two thrust bearings mentioned a variety of other designs have been developed, but for the most part they differ in features of application and adjustment rather than in principle.

Tandem-geared Machines. On tandem-geared machines (Fig. 1-9, page 14) the gears not only mesh into the worm shaft, but they mesh into each other, as can be seen in Fig. 1-9. In this arrangement the thrust is taken in the gears and worm shaft and no thrust bearings are required. This arrangement works well as long as the gears do not wear, but as soon as these parts begin to wear, backlash in the gears permits end motion of the worm shaft. There is no way that this end motion can be corrected except by installing thrust bearings or renewing the gears, the latter being an expensive job.

Worm and Gear. Don't work on the worm shaft or thrust bearings until the car or counterweight, whichever is heavier, is bottomed in the hoistway. On modern machines this will always be the counterweight. If the gear is allowed to get out of mesh with the worm, the counterweight, being heaviest, will fall, with disastrous results. When work is to be done on drum-shaft bearings support the counterweight from the bottom of the hoistway and sling the car from the overhead beams, using two chain falls.

When the gear and worm are not properly lubricated, they may operate with a loud roaring noise and vibration. This problem has been solved by putting one or two pounds of sulfur in the gear case with the oil and letting it remain until gear operation becomes normal. The length of time will depend upon the elevator service. If the machine is in continuous operation, one-half to one day is sufficient to put a good wearing surface on the gear and worm. Then the oil should be removed from the gear case, and the gear case thoroughly cleaned before new oil is put in. Chipped balls in a thrust bearing will cause the machine to make a noise similar to that made when the worm and gear are cutting, but nothing will cure the thrust-bearing trouble except renewing the defective parts.

When it was general practice to key worm gear and drum to the drum shaft as separate units, considerable trouble was experienced with these parts working loose on the shaft. In modern practice the drum is keyed to the shaft, and the gear bolted to a heavy flange extension on the drum's hub. If the bolts are properly tightened and the nuts locked in place, loose gears are a rare occurrence. However, during inspections, this part of the machine should be carefully checked, since a failure here would be dangerous, as the car would be free to go in whichever direction excess weight in the counterweight or car would take it.

Wearing of Traction-sheave Grooves. On traction machines, either of the geared type (Fig. 13-2) or direct type (Fig. 13-4), wearing of grooves in the traction sheave to different diameters is a source of trouble, as it affects the life of the ropes. Where the grooves have unequal diameters, the ropes on the smaller-diameter grooves must slip to have the same speed as those running on the larger-diameter grooves. This slipping tends to increase the wear and make conditions worse. Not only are the grooves worn, but the ropes will wear to different diameters.

When the grooves become worn, it is now general practice to mount a lathe tool on the elevator machine and condition the grooves,

Fig. 13-4. Direct-traction elevator machine with brake mounted on the motor.

using the motor to drive the sheave. This requires shutting down the machine and removing the ropes from the sheave. Turning the grooves to the same size and diameter does not completely remove the trouble. Owing to the ropes being of unequal diameters, on V-groove sheaves the smaller ropes will go down deepest in the grooves and there will be a differential wear between the ropes and grooves. Therefore, when the grooves are trued up the ropes should be renewed.

As already mentioned, the mechanical faults that may develop in an elevator machine will vary somewhat with the type of equipment.

But the foregoing will act as a guide to the most common defects. In most cases the mechanical defects that develop on elevators are quite obvious to a good mechanic. This is not always true with electrical troubles. Although the mechanical difficulties do not vary widely on the different classes of machines, there is a great difference in the electrical faults that may develop.

CHAPTER 14

Locating Faults in Direct-current Motors and Controllers

Classification of Faults. Faults in electric elevators may be divided into four classes: 1, power supply; 2, motor; 3, control equipment; and 4, mechanical. On account of the wide variation in the power supply, types of motors and controllers, and the design of mechanical details, an almost infinite number of troubles can occur in elevator equipment. This does not mean, however, that electric elevators, when properly designed, installed, and maintained, give any more trouble than other similar apparatus. This chapter indicates in a general way how to go about locating the cause of trouble when it occurs.

Faults Due to Power Supply. If an elevator is to operate satisfactorily, the power supply must be correct for the motor and control equipment. Low voltage on a d-c system causes motors to run below rated speed and may cause controllers to function improperly. On the other hand, high voltage increases motor speed above normal and may overheat the field and magnet coils. However, the latter is not likely, on account of the intermittent service, but on elevators the power supply must be up to par.

Motor Troubles. D-c elevator motors are of the shunt or compound types. On direct-traction machines (Fig. 14-1), a shunt motor is generally used, operating at speeds of about 60 to 125 rpm. For geared traction machines (Fig. 14-2), and the drum type (Fig. 13-1), compound motors operating at a comparatively high speed are used.

As elevator motors in their general design are not different from those found in other industrial services, they are subject to about the same troubles. In d-c motors these troubles may be divided into: sparking at the brushes, heating, excessive speed, too low speed, noise, and failure to start. With some of the earlier types of motors that were not designed for reversing service, it was impos-

251

sible to locate the brushes in a position to give sparkless operation on the commutator, because this position changed from one direction of rotation to reverse. For reversing service the motor's brushes must be set on the neutral. With modern types of machines, if other features of operation are correct, the motor will operate sparklessly with the brushes on the neutral.

On account of frequent starting and stopping, the controller must be adjusted so as not to accelerate the motor too rapidly, or the

FIG. 14-1.   Direct-traction elevator machine.

brushes may spark each time the motor starts, which will gradually roughen the commutator and cause trouble at the brushes. On most elevator controllers the motor armature is connected through a resistor during stopping for dynamic braking. If the dynamic-braking resistance is of too low a value, its current will be excessive and cause sparking at the brushes during stopping. Even though this sparking is only for short periods, it will gradually roughen the commutator and cause sparking at the brushes during the full operating period.

Other causes of sparking at the brushes are: brushes not set diametrically opposite; brushes not in line with the commutator bars; brushes not in good contact with the commutator, due either to dirt and oil on the brushes or to the condition of the commutator; high mica in the commutator; open circuits or short circuits in the armature winding; an open circuit or short circuit in the field coils; and wrong polarity of field coils.

Heating of an elevator motor is not so likely to occur as in industrial applications where the load may be increased to overload values. It is seldom, if ever, that an elevator motor is overloaded long enough to cause serious heating. The brake, even if it were so adjusted as to cause overloading of the motor, would become so hot as to call attention to the condition before the motor was injured. The chief causes of overheating of an elevator motor are:

Fig. 14-2. Heavy duty geared traction-elevator machine with two brakes, one being on the traction sheave.

short circuits in the armature or field coils, moisture in the windings, the windings becoming oil soaked, or the bearings worn so as to allow the armature to rub on the pole pieces.

Because of the nature of the load it is seldom that trouble is experienced on the motor with hot bearings if they are kept properly lubricated with a good grade of oil. On geared-type machines with disk-type thrust bearings it is possible that these bearings may wear sufficiently to allow shaft thrust to be taken on one of the motor bearings, in which case this bearing may overheat.

Excessive Speed. The diagram in Fig. 14-3 shows current direction through a compound machine as a motor. Here the current in the series- and that in the shunt-field coils are in the same direction, which is correct. In Fig. 14-4 it is assumed that the motor's speed has been increased, by the elevator car and its load in down direction, to where it has become a generator and is pumping back into the power system. In this case, current direction in the armature and series-field winding is reversed. Series- and shunt-field windings now have opposite polarity to weaken the field poles, and they tend to reduce the braking current and let the elevator increase in speed. To overcome this difficulty, the controller is so arranged that the series-field winding is in circuit only while the motor is accelerating, after which it is cut out of circuit and the motor operates as a shunt machine. Should the controller not cut out the series winding, the motor will overspeed with a heavy load in the car in down motion. It may also overspeed in the up motion with light load in the car and heavy counterweight.

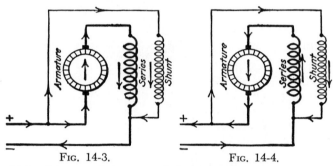

Fig. 14-3.   Fig. 14-4.

Fig. 14-3.   Diagram of a compound elevator motor operating as a motor.
Fig. 14-4.   Diagram of a compound elevator motor operating as a generator.

In the down direction with a heavy load in the car, if the armature resistance is not cut out of circuit, there will be a tendency for the car to overspeed. On account of the added resistance in the armature circuit, the motor operating as a generator has to run at an increased speed to supply the necessary braking current to hold speed near normal.

Open circuits or short circuits in the shunt-field windings will also cause excessive speed, but will generally blow the fuses, when the series-field winding is cut out of circuit. Shifting the brushes back of the neutral will cause an increase in speed, but it can also produce severe sparking which may cause flashing over on the commutator and opening of the fuses or circuit breaker.

Causes of Fuses Blowing.   Other faults in the motor that may blow fuses are: short circuit of a number of coils in the armature; open circuits in the armature, where they produce sparking severe enough to cause flashing over on the commutator; open circuit in the shunt-field coils; short circuit between the shunt- and series-field coils; and grounds in the field coils or armature windings. In some cases the brushes may start to chatter and cause a flash-over on the commutator, resulting in fuses blowing.

Another cause of the fuses blowing is the connection of the series- and shunt-field windings in opposite polarity. On account of the comparatively large number of turns used in the series winding of compound motors for elevator service, it will generally be impossible to put the motor into service with the field coils in opposition without causing excessive speed or blowing of the fuses. However, if trouble is experienced with a compound elevator motor, the polarity of the field coils should be checked.

Speed Too Low. Causes of the motor failing to come to speed are usually failure of the controller to function properly, such as not cutting out the starting resistance; brake not releasing; etc.; and will be discussed later in this chapter. The armature rubbing on pole pieces, due either to worn bearings or to a loose pole piece, would cause a reduction in speed. A short circuit of a number of coils in the armature is another cause of motors failing to come up to speed. Any of these things will generally cause the motor to overheat.

Locating Troubles in Controllers. If the electrical equipment of an elevator is causing trouble, its action will give the trained trouble man a good idea where the fault is. For example, let us assume that the motor fails to start. The first assumption here would be that the main fuses are blown or the power is off the system, and these are the first things to be checked. If the controller is of the type with a potential switch, this switch remaining closed generally indicates that the circuit is alive, since the potential switch remains closed as long as power is on the line and the emergency limits are closed.

When the fuses are blown, they should be replaced and an investigation made for the cause. Inspect all controller contactors to see that they are in normal off position and in good operating condition. Check brush position on the motor to see that the brush yoke is locked in position. If the motor and controller are found in good condition, test for ground. When making this test, do so not only on the motor, but also on all terminals on the control board.

Testing of Motor Shunt-field Winding. After testing for grounds and finding motor and control clear, test the shunt-field circuit to see that it is complete. How to make this test depends on how the field winding is connected. If the shunt-field winding is connected directly to the line, as in many types of controllers, closing the line and potential switches will energize the field winding. If the field circuit is complete, the line switch will arc when it is opened, and the field poles will have a strong attraction for a screw-driver or a pair of pliers if placed against one of them when the coils are energized.

On some controllers the shunt-field winding is disconnected from the line with the rest of the motor every time the elevator is stopped. In these cases the field winding can be tested by opening one side of the armature circuit, then closing the line and potential switches and throwing the control in the car to either the up or the down position. This will allow the controller to function and energize the field coils with the armature circuit dead. Indications of the field circuit being complete are as previously mentioned. When either one of these field-coil tests is made, current in the coils can be limited by connecting a lamp in series with them. If the circuit is complete, the lamp will light.

After these tests have been made on the motor and controller and no cause of the fuses blowing is found, or if a cause has been found and removed, the line and potential switches may be closed and the elevator started from the car switch. If it runs all right, it should be watched for a few trips to see if anything irregular develops in the control.

Precautions Taken before Starting Elevator. Before starting the elevator, block the accelerating switches open so that the starting resistance cannot be cut out. In this way a test can be made for an open circuit in the starting resistance. With the accelerating switches blocked open the motor should start, if the starting resistance and the rest of the armature circuit are completed.

Give particular attention to the motor when it is first started, to see that it functions properly. A group of armature coils short-circuited will prevent the motor from coming up to speed and cause it to act as though it were heavily loaded and to heat.

When the elevator is started with the accelerating switch blocked open, watch the motor and starting resistance for anything that might cause the latter to overheat. For example, a short circuit in a group of coils in the armature would cause it to take a heavy current through the starting resistance and cause this resistance to heat excessively.

Also watch the brake to see that it releases properly. After the motor operates correctly with the starting resistance in circuit, release the accelerating switch and bring the elevator to full speed, watching the motor and control at all times for any irregularities.

Causes of Potential Switch Not Staying Closed. Assume a condition where the fuses are found to be all right, but the potential switch will not remain closed. This condition would indicate an open in the holding-coil circuit of this switch, in which various protective devices are connected in series with the coil. On a simple hand-rope-controlled drum-type machine, these may consist, as shown in Fig. 14-5, of a slack-rope switch *SC*, governor switch *GS*, upper hoistway-limit switch *UL*, lower hoistway-limit switch *LL*, safety switch *S* in the car, and on some machines a final drum-shaft limit

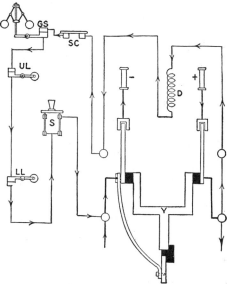

FIG. 14-5. Diagram of potential-switch holding-coil circuit for a slow-speed elevator.

switch. Opening any of these devices will deenergize holding coil *D* and allow the potential switch to open.

For example, if the car were to go down on the bumpers and the ropes became slack, the potential-switch-coil circuit would be opened by the slack-rope switch and the lower hoistway-limit switch. It would be necessary in such a case to take up the rope slack on the drum and raise the car off the lower hoistway-limit switch before the potential switch would remain closed.

After the machine has been moved by hand until the slack is taken out of the ropes, the car can be raised by holding the potential switch closed and operating the machine from the hand rope or car switch, or by closing the up-motion direction switch on the controller by hand. Care must be exercised in doing this, as a mistake may cause the counterweights to be pulled into the overhead work, with disastrous results.

Car or Counterweights Landed. When the car or counterweights are bottomed and the ropes are slack on the drum, do not try moving the machine with the motor, as there is danger of getting the ropes crossed and of subjecting them to heavy stresses. With the car ropes slack the total weight of the drum counterweight is effective in turning the drum. If the brake is released and the machine started by the motor, the slack is likely to be taken out of the ropes more quickly than is good for the equipment before the brake can be applied again.

If an attempt is to be made to move the machine by the motor, open the brake-coil circuit so that the brake cannot be released and then close the up-direction switch by hand at the control board, if the car is bottomed. In this case the brake remains applied; if the motor moves the car, it will do so very slowly, and just as quickly as the power is cut out of the motor it will stop.

The same operation can generally be done to better advantage by using a spanner wrench on the brake-wheel coupling. Where this is not possible, an iron bar can generally be used in the brake wheel as a lever to turn it in the desired direction. This turning can usu-ally be done without releasing the brake. On some machines the motor shaft extends beyond the outboard bearing, with a square end to allow using a wrench to move the machine by hand.

On elevators having brakes applied by a weight on a lever, one man on the brake lever to control the motion of the car and another to turn the brake wheel by hand make for safety in taking slack out of the ropes. Where the motor shaft is extended for use of a wrench to move the elevator by hand, do not use a Stillson wrench, since after the machine starts effort may have to be applied to the wrench in the reverse direction to control machine speed. There is danger of a Stillson wrench opening sufficiently to lose hold on the shaft.

Potential-switch-coil Circuit on High-speed Elevators. On high- or medium-speed traction elevators the potential-switch circuit becomes somewhat more involved than on drum-type elevators or slow-speed traction machines. Figure 14-6 shows the potential switch for a high-speed traction elevator. Starting from the $+$ side of the line the circuit is through the single-pole swith S_p and a fuse on the control board. From the control board the circuit goes through contacts L on the car-governor switch, then through contact $+L$ on the counterweight governor, contact $+L''$ on the top hoistway-limit switch, contact $+P$ on the bottom hoistway-limit switch, and the slack-rope switch on the compensating-sheave frame at the bottom of the hoistway, then back to the control board and through coil L of the potential switch. When this switch is open, contacts A_l are closed and resistance ARP is cut out of circuit with coil L, so that when the coil is energized it can close the switch. Closing the potential switch opens contacts A_l and cuts resistance ARP in series with coil L to reduce its current.

From the control board the circuit continues through contact O_m on the overload relay, contact $-P$ on the top hoistway-limit switch, $-L'$ on the bottom hoistway-limit switch, one side of the emergency switch Y in the car, and the safety switch Y' under the car. From the car safety switch, the circuit is back to the control board, through

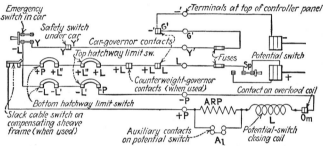

Fɪɢ. 14-6. Diagram of potential-switch closing-coil circuit for a high-speed traction elevator.

a fuse to contact G' on the car governor switch, and then back to the — main contact of the potential switch on the control board.

Opening any of these devices in the potential-switch circuit will cause it to open and remain open as long as an auxiliary switch is open. Limit switches, located in the elevator hoistway or under the car, are subjected to dirt, grease, and moisture in varying degrees. Therefore, their contacts should be given careful attention and kept clean, to prevent grounds and short circuits and to keep dirt from working into the contacts, interrupting the potential-switch-coil circuit and shutting the elevator down.

Locating Cause of Potential Switch Opening. When the potential switch will not remain closed, and it has been ascertained that the line is alive, clean all devices in the closing-coil circuit to make sure that they are closed and making contact. Do this after testing the circuit fuses to determine if they are complete. If the fault is not found in any of these auxiliary switches, then the circuit through the coil should be tested to see if it is open. After this has been done and the coil found in good condition, it will then be a case of testing the wiring. The most likely place for the wiring to open is in the traveling cable from the junction box, located halfway up the hoistway, to the bottom of the car.

A broken wire in a traveling cable may be difficult to locate because as the cable bends, the wire may open-circuit at one location of the car and close at another, or the wire may be open and a slight movement of the cable may cause it to close. If the potential switch remains closed at times and not at others it is a pretty safe indication that the fault is in the traveling cable and it should be renewed, unless there is an extra wire in it.

When trouble of this kind develops, run the car to just above the control-circuit junction box in the hoistway, where the traveling cable can be pulled out on the floor. Then with the potential switch closed, begin at one end of the cable and gradually work along it, bending the cable back and forth. In this way the broken wire will generally open and the potential switch will drop out. Station a man at the control board to note when this happens. If the fault in the cable can be located, it can be repaired, but this is generally not a recommended course, except as a temporary measure, since the old cable is probably in such condition that trouble may develop in a short time at another place.

Switch S_p (Fig. 14-6) has two positions. When closed to the right as in Fig. 14-6, the potential switch will close when its coil circuit is energized. When switch S_p is thrown to the left, the potential switch must be closed by hand, after which it will remain closed if the coil circuit is energized. These two arrangements will cause different effects if resistance ARP, in series with the potential switch coil L, is open.

The foregoing indicates in a general way what may cause the main fuse to blow or the potential switch to open. These conditions will vary widely with different controllers. On some equipments the potential switch coils are in series with the direction switch coils, so that the potential switch opens and closes with the direction switches. In such a connection almost anything that would affect the closing of the potential switches would also affect the direction switches.

Testing the Direction Switches. After it has been ascertained that conditions in the power circuit are correct for operating the elevator motor and controller, as previously described, and the motor fails to start, then give attention to the other circuits. On elevators that are operated from a hand rope or a wheel in the car, the direction switches are usually of a mechanical form, such as a cylindrical design, or a type closed and opened by cams on the shipper wheel, as shown on the right of Fig. 13-1, or any of numerous other devices that are employed. These switches, being of the mechanical type, are positive in their action and should close when the control mechanism is thrown to either the up or down direction.

Assume that it has been found that the power circuit is complete and alive to the controller, and the motor fails to start when the direction switch is closed. There are two possible conditions. One is where the motor will not start with the switch closed for one direction and will start with the reverse switch closed for the other direction. Such a condition indicates that the trouble is in the reverse switch, which, when closed, prevents the motor from starting. This confines the search for the trouble to a small part of the equipment. In such cases it will generally be found that the contacts on the switch have become worn or are out of adjustment and do not touch to make the circuit. It is also possible that the contacts have become broken or otherwise made inoperative, or that their connections have become loose or broken.

To inspect the reversing switch, open the line switch and pull the control to the position at which the motor failed to start. It then can be readily seen how the contacts fit, and loose or broken connections can be looked for at the same time.

The second condition is where the motor fails to start with the reverse switch thrown for either direction. It is possible for the reverse switch to be out of order for one direction, but not very likely for both directions at the same time. Therefore, if the motor fails to respond with the reverse switches in either direction, it is a pretty good indication that the cause of the trouble is somewhere else than in the switches. The first thing to observe is that the switches function properly, as this will show whether the fault is in the mechanical operating mechanism or in the electric circuits.

Assume that the direction switches have been found to function properly. Then the source of the fault may be assumed to be in the motor-circuit wiring. To make sure that the power circuit is complete to the controller, it is well to make a test across the first two line terminals in the controller. For example, in Fig. 14-7 the + side of the main line or potential switch *PC* is connected to 5 on the direction switches, and the − side is connected to *H* on the controller. If a lamp is connected from 5 to *H* and the potential switch closed, the lamp should light. Failure of the lamp to light indicates that something is wrong in the power circuit before it actually reaches the controller.

Where only one terminal can be found on the controller, one lamp lead can be connected to that and the other tested on the part of the controller to which the opposite side of the power circuit leads. For example, in Fig. 14-7 only terminal *H* is on the controller, 5 being on the reverse switches located on the end of the drum shaft. In this case one lead of the test lamp can be connected to *H* terminal on the controller and the other lead tested on the reverse switch located over the end of the drum shaft. If the circuit is complete, a light should be obtained. If the lamp does not light, the fault must be located somewhere between where the test was made and the power supply.

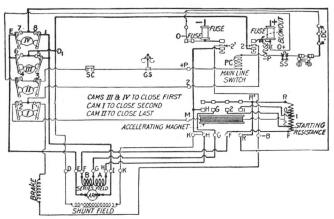

Fig. 14-7.   Diagram of a semimagnetic controller for drum-type elevator.

Open Circuit in Starting Resistance.   When the test shows that the circuit is alive to the controller and the motor will not start with the direction switch in either position, the most likely place to find the fault is in the motor or starting resistance, although it is possible for the fault to be in any part of the armature circuit. The starting resistance and series field may be tested by connecting a lamp in series with the armature, then closing the line, direction, and accelerating switches by hand. Closing the accelerating switch cuts out the starting resistance and series-field winding. Therefore the lamp should light if the armature circuit is complete. In this way the trouble source is isolated and can be easily located.

On some machines with mechanical-type controls the motor circuit goes through an ultimate limit switch that is opened by the drum-shaft limits when the car goes by the terminal landings. These switches have been known to open accidently; so when the motor fails to start it should be ascertained if such a limit switch is part of the control equipment and whether or not it is in the closed position.

A more likely cause of the motor not starting properly is a short circuit in a group of coils in the armature. In this case the armature will generally start, but will not come up to full speed, rather turning slowly and acting as if heavily loaded. This will cause heating of armature coils and starting resistance and may blow the main fuses if the controller is of a type that will cut out the starting resistance under overload conditions.

Another cause of the motor failing to start is the shifting of the brushes off the neutral; their position should be checked.

Where power is supplied from a two-wire ungrounded system, if a ground occurs on the side of the line going directly to the motor and also in the armature or series winding, the motor will not start, and if part of the starting resistance is cut out, the fuses will blow. If the ground is on the side of the line in which the starting resistance is connected, the fuse will blow as soon as the reversing switch is closed. On a three-wire system with a grounded neutral the fuses will blow whenever a ground occurs in the armature circuit of the motor, operating on full voltage, and the reverse switch is closed. Combinations of grounds in the motor and controller have been known to allow the elevator to operate in one direction and to blow the fuses when the reverse switch was thrown to the other direction.

Testing of Shunt-field Winding. With an open in the shunt-field circuit, the motor will generally start if it is not too heavily loaded, but the fuses will blow if the starting resistance and series-field winding are cut out of circuit. A test can be made for an open in the shunt-field-coil circuit by opening the armature circuit and pulling the controller to the on position. For example, in Fig. 14-7 the armature leads can be disconnected at *I* or *E*. When this is done, precautions should be taken to prevent the loose lead from making contact with some other part of the motor and causing a short circuit when the controller is pulled to the on position.

The brushes may be insulated from the commutator by using a piece of cardboard instead of disconnecting the armature leads. When this is done, care should be exercised to see that carbon and copper dust on the brushes do not form a conducting path from brushes to commutator and cause an arc that will damage the commutator when the switch is closed. When the controller is put in the on position with the armature circuit open, if the field coils are energized the pole pieces will attract a screw-driver or any piece of iron brought near them. When the switch is opened, an arc will be obtained at its contacts.

Brake Fails to Release. The brake not releasing may prevent the motor from starting when the circuit is complete through it. Where the brake is mechanically operated, it is not likely to fail to open. As the lining wears in the brake shoes, adjustments are generally made to take out the increased clearance between brake shoes and pulley when the brake is released. When old linings are replaced with new ones, the latter may be so thick that the operating mecha-

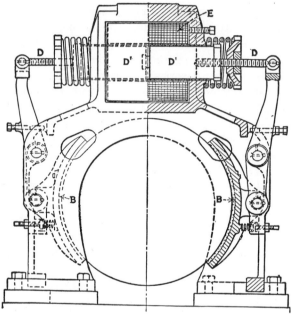

Fig. 14-8.   Magnet-operated brake with single coil and two cores.

nism cannot lift the shoes far enough to release the brake until it has been adjusted for the new linings.

When the brake is operated electrically, there are a number of causes for its failing to release. First, the coil circuit may be open, due to a broken wire in some part of the circuit, or the contacts that open and close the circuit may be out of adjustment, worn so that they do not close, or they may be prevented from closing by dirt between them. If the circuit is complete, the cores may be so far out of the coil that the latter will not develop sufficient pull to release the brake. For example, in Fig. 14-8, the more the brake-shoe linings B wear the farther the cores D' are pulled out of coil E when the brake is applied, and it is possible for them to be far enough apart to reduce the pull of the coil to the point at which the brake will not release.

Adjustments are made by screwing the stems *D* in or out of the cores so that when the cores pull together the brake shoes will be just lifted clear of the brake wheel. As the brake lining wears, stems *D* should be backed out of the cores to maintain the adjustment. If such an adjustment is maintained until the brake lining is worn to the point at which it has to be renewed, when the new lining is put into the shoes with the old adjustments, the cores will touch when the brake shoes are in contact with the pulley. To correct this condition stems *D* should be screwed into cores *D'*, separating the latter enough to give sufficient movement to lift the brake shoes clear of the pulley.

On some brakes a resistance is connected in series with the brake coil. This resistance is short-circuited by contacts when the brake is applied. This allows connecting the coil directly across the line to obtain a strong pull for releasing the brake. When the brake releases, contacts short-circuiting the resistance open and cut it in series with the brake coil. If these contacts get out of adjustments and do not close when the brake is applied, the resistance will remain in series with the coil and it may not develop sufficient pull to lift the brake shoes.

On the other hand, if the resistance is open-circuited and the contacts close, as they should when the brake is applied, this will cut the resistance out of circuit. When the brake coil is energized, it will start to release the brake; but when the resistance contacts part, the coil circuit will be opened and the brake immediately applied again. In some cases the arc at the contacts, when they open, may hold across and the brake remain released. In such an event the contacts will soon burn so short that they will not meet when the brake is applied. The circuit will then be opened through the resistance, and the coil will not be able to release the brake.

Accelerating Switches Fail to Function. Failure of accelerating switches to function and cut out the starting resistance will depend largely upon the principle on which these switches operate and on how they are connected in the circuit. In any case an open circuit through their coil or coils will prevent operation. This open may be in the wiring, it may be caused by a loose connection, or the auxiliary contact that closes the coil circuit may be out of adjustment or make a poor connection.

On the controller shown in Fig. 14-7, the accelerating switch operates on the counterelectromotive-force principle. In this case the motor must start and build up a certain voltage across its terminal before sufficient current flows in the coil to cause contactors 1, 2, *G*, and *H* to close and short-circuit the starting resistance. If the motor is too heavily loaded, it may not be able to reach the speed at which the contactors will close. Adjustments are made by screws; when the screws are raised or lowered, the arms come closer or farther away from the magnet core. Adjustment can also be made by shifting connection *O*, shown in Fig. 14-7, where the coil connects into the starting resistance. Moving connection *O* toward *F* tends

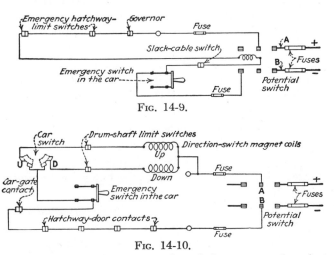

Fig. 14-9.

Fig. 14-10.

Figs. 14-9 (above) and 14-10. Potential- and direction-switch coil circuits on drum-type elevator controllers.

to make the contactors close at a lower speed on the motor with a given load, whereas moving *O* toward *R* has the opposite effect. When the elevator is installed, the proper location of *O* is made by the manufacturer and the only subsequent adjustments that may have to be made are of the contactors.

Methods of adjusting the accelerating switches vary with the type and make. If they are adjusted so that they accelerate the motor too quickly, the fuses may blow or an unpleasant motion may be imparted to the car when it starts.

Full-magnet Controllers. On a full-magnet elevator controller, limit and safety switches are divided between the potential-switch and direction-switch coil circuits. Figures 14-9 to 14-13 show some common arrangements of potential-switch and direction-switch circuits. Figure 14-9 shows the potential-switch circuit for a drum-type elevator; in addition to the other devices it has a slack-rope switch, which will be opened should the car or counterweights land on the hoistway bottom and the ropes become slack.

On a traction-type machine the ropes cannot become slack; so the slack-rope switch is replaced by a switch under the car, which opens if the governor sets the safeties. In Fig. 14-11 the hoistway-limit switches are shown double-pole, as in Fig. 14-14; while in Fig. 14-9 they are single-pole, as in Fig. 14-15. These switches are mounted at the top and bottom of the hoistway and are opened by a cam on the car striking roller *R*.

On some drum-type installations both hoistway-limit switches in the potential-switch circuit are mounted at the bottom of the hoist-way, so that one is opened by the car and the other by the counter-weights. The latter occurs if the car overtravels the top landing. As has been previously discussed, if any device in the potential-switch-coil circuit is not making proper contact, it will cause the potential switch to open. In all cases of trouble involving the elevator's failure to start, make sure that the power circuit is complete up to the potential switch.

Testing the Power Circuit. Too much emphasis cannot be laid on being sure that the power circuit is alive and complete to the controller, and the only sure test is that made with a lamp or voltmeter connected across the line. On a three-wire 110- and 220-volt circuit with grounded neutral, tests for power on the 220-volt circuit are sometimes made with a 110-volt lamp, connected one side to ground. That is, one side of the lamp is grounded and the other lead is connected first to one side of the circuit and then to the other. If a light is obtained on both tests, it is assumed that the power circuit is all right, but such a test is not always reliable.

Testing of Direction-switch Control Circuits. When the potential switch remains closed and the motor fails to start after the control switch in the car is put in the on position, attention should be given to the direction-switch circuits. There are two conditions that will generally be met in a car-switch control. That is, when the car switch is placed in the on position, the direction switch on the controller will either close or fail to close.

From either of these conditions a fair idea may be formed as to where the trouble may be. If the direction switch responds to the control switch in the car, it is known that this circuit is not only alive but is in operating condition. If the direction switch closes and the motor does not start, it will indicate that the trouble is in the motor circuit itself. If the motor will not start with either one of the direction switches closed, it is a pretty sure indication that the trouble is not in the direction switches, but in other parts of the motor circuit. It is possible that the circuit is open in one switch, but the motor circuits in both switches are seldom open at the same time. Therefore, if the motor fails to start with either reverse

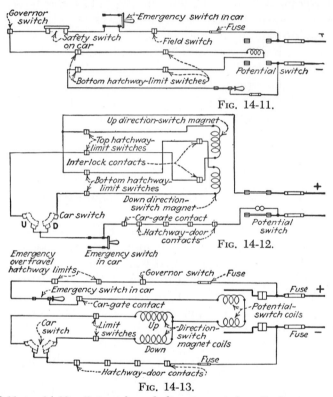

Fig. 14-11.

Fig. 14-12.

Fig. 14-13.

Figs. 14-11 to 14-13. Potential- and direction-switch coil circuits on traction-type elevator controllers.

switch closed, the possibilities of the trouble being in these switches are remote.

If the direction switches are eliminated as the source of trouble, the most likely sources are in the motor or in the starting resistance. The starting resistance may be tested by connecting a lamp in series with the armature and closing all accelerating switches; this will cut all the resistance out of circuit, and if the trouble is in this part of the circuit the lamp will light. These tests were described more fully on page 265.

If the direction switches fail to close, it is an indication that their coil circuits are not complete. Figure 14-10 shows the direction-switch circuits for a drum-type machine. Starting at *A* on the potential switch, the circuit is through a fuse on the control board, then through either the up- or down-motion direction-switch magnet coil, depending on the position of the car switch. From the direction-switch magnet coil the circuit continues through the drum-shaft limit switches, the car switch, one side of the emergency switch in the car, the car-gate contact, and the hoistway-door contacts, to *B* on the — side of the line.

Fig. 14-14.   Fig. 14-15.

Figs. 14-14 (left) and 14-15. Double-pole and single-pole hoistway-limit switches.

Any of these devices not making contact will prevent the direction switches from closing. It is possible that one switch will close and the other will not, at locations other than the terminal landings. For example, if the down-direction-switch coil circuit is open between the coil and the car switch, this switch will not close when the car switch is in the down position, but this will not interfere with the circuit of the up-direction-switch coil. On the other hand, if the open is between the car switch and the — side of the line, it will interrupt the circuit of both direction-switch coils and neither will respond to the car switch. These two conditions give an idea of whether the trouble is in the individual coil circuit or in that part of the circuit that is common to both coils.

If neither direction switch responds to the car switch, a test can be made to determine if the trouble is in the direction-switch coil circuit or in the power circuit by closing one of the direction switches by hand. With the power circuit complete the motor should start when the direction switch is closed by hand, or at least an arc should be obtained at the contacts when the switch is allowed to open. The absence of these indications show that power is not getting to the controller.

When either of the direction switches is closed by hand, care must be exercised to be sure that it is safe to move the car and that the car is not run by the terminal landings. Unless the equipment is protected by emergency limits and a potential switch, there is danger of pulling the car or counterweights on a drum-type machine into the overhead work, with disastrous results.

On some controllers an operating switch is provided on the control board for operating the car from this point. This switch is generally connected in the circuit so as to cut out all safety switches in the circuit with the exception of the terminal floor limits, and it can operate the car on the first speed only. If the car will operate from the control-board switch and will not operate from the car switch, it shows that the trouble is in the part of the circuit not included in the circuit of the former. This information assists in narrowing down the number of places to look for the trouble.

Assume a case in which there is no operating switch on the control board; the motor starts when the direction switches are closed by hand, but neither of these switches respond to the car switch. One of the first things to do is to examine all contacts in the car-switch circuit. These will vary with the type of installation and the locality. In Figs. 14-10, 14-12, and 14-13, gate and door contacts are shown. On some old installations these are not used, but when they are, they should be one of the first trouble sources looked into. On cars provided with an emergency switch for short-circuiting the door contacts, the test for the trouble in these contacts can be made by closing this switch. If the car can be operated when the door contacts are out of circuit and cannot when they are in circuit, some of these contacts are out of order.

Unless the car switch is given careful attention, the contacts may become worn or out of adjustment and fail to complete the circuit. If the trouble cannot be found in any of the limit and safety switches, attention should be given to the wiring. If there is a fuse in this circuit, it should be one of the first things tested. Owing to the bending of the traveling cable, which makes the connections between the hoistway wall and the car, the cable is always a potential source of trouble. When the trouble is in the traveling cable, it is likely to disappear temporarily if the car is moved by closing the direction switches on the control board by hand. When trouble of this kind is experienced, renew the traveling cable, as locating the open circuit and repairing it usually gives only temporary relief, as previously explained.

Direction switches are generally equipped with an interlock so that one switch must be open before the other can close. When this interlock is electrical, it consists of contacts on the switches through which the coil circuits are made. The contact for the down-direction-switch coil is closed when that for the up switch is open and vice versa. These contacts are indicated on Fig. 14-12. On such switches unless the interlock contact on one switch completes the circuit, the other switch cannot be closed from the car switch. A mechanical interlock consists of a bar fulcrumed in the center so that when one switch closes, it holds the other open.

On some types of controllers it is possible for the direction switches to get out of adjustment so that their magnet coils cannot close them. On other controllers, using horizontal solenoids for closing the reversing switches, their movable cores wear into the brass sleeves of the coils to prevent the switches from closing properly. On such switches dirt is liable to affect the cores and sleeves adversely and prevent the switches from returning to the off position to close the interlock contacts, thus causing unsatisfactory operation of the elevator. In other cases the contactors are made sluggish in their operation by the lack of lubrication on the bearings that support the contactor arms. A little lubrication on the bearings during the regular inspections will help to ensure satisfactory operation of the contactors. Grounds in the wiring or short circuits in the direction switch coils will cause the fuses to open in these circuits and prevent operation of the switches.

Direction-switch coil circuits discussed are essentially the same on all elevators, although there may be considerable differences in the safety devices included in these circuits. On a drum-type machine, terminal limits will be on the end of the drum shaft, whereas on a traction-type machine, these switches may be mounted in the hoistway, near the terminal landings, and opened by a cam on the car. On traction machines limit switches may be mounted on top of the car, in which case they are opened by a cam in the hoistway. In some installations one or more switches are on the governor (Fig. 3-11, page 89); in other cases such switches may be omitted.

D-c elevator controllers are designed for a wide variety of speeds. For the slow-speed machines only one speed is used, in which case there are generally four contacts on the car switch. On a two-speed controller there are six contacts; the two top ones give the second speed. On high-speed machines, as many as six different speeds may be obtained, in which case the car switch will have as many as 14 contacts. However, in any case, the first contact to make in either direction when the car switch is moved to the on position is the direction-switch circuit.

On machines that operate at more than one speed the cause of the machine's failing to come up to full speed after starting depends on the type and make of controller, and a careful study of high-speed elevator-control circuits is to be recommended. However, if the various contactors on the controller and the car switch and limit switches are kept in good condition, about 90 per cent of the causes of trouble will be eliminated. This also applies to push-button-controlled automatic elevators.

CHAPTER 15

Locating Faults in Alternating-current Motors and Controllers

Types of A-C Motors. Faults in a-c motors used in elevator service will be the same as those in these motors when they are used for other applications, except in so far as the motor's operation may be affected by the conditions of the elevator machinery and the controller. With direct current, shunt or compound motors are used for elevator service. In a-c elevator practice a wider diversity of motors is used and this somewhat complicates the problem of locating troubles when they do occur. Although single-phase-type motors have been used to drive elevator machines, these are exceptions to general practice. Therefore only the polyphase types will be considered.

On a-c circuits polyphase motors of both the squirrel-cage type (Fig. 15-1) and in some cases the wound-rotor type (Fig. 15-2) will be found in elevator service. For machines with speeds up to about 100 fpm the simple squirel-cage motor is generally used. For higher-speed drum-type or geared traction machines (Fig. 15-3) multispeed motors are used. These motors may have two windings in the same slots, each winding grouped for a different number of poles; for example, 8 and 4 poles or 12 and 4 poles, giving a speed ratio of 2 to 1 or 3 to 1, as the case may be. In some cases, only one winding is used, and this is regrouped for two different numbers of poles by the control equipment.

Two different motors have also been used. The stators of the two may be mounted in the same frame and the rotors keyed to the same shaft so that the two motors form a single unit. All starting and accelerating is generally done on the high-speed winding, the slow-speed winding being used for slowdown and stopping.

Because proper speed control cannot be obtained, and there are

273

difficulties in designing and building very slow-speed a-c motors, direct-traction a-c elevator equipment is not used.

Classification of Faults. Even though conditions under which troubles occur on the motors vary somewhat with the different types, practically the same difficulties may develop in all types. These troubles may be divided into: noisy operation; motor fails to start; torque too low; speed too low; speed too high; heating; sparking at the brushes on slip rings, where the motor is of the wound-rotor type; and reverse rotation.

With alternating current, the quality of the power supply is affected not only by the voltage, but also by the frequency of the system. Torque of an a-c motor varies as the voltage squared;

Fig. 15-1. Drum-type elevator machine driven by single-speed squirrel-cage motor.

therefore a comparatively small variation in voltage may materially affect the torque of the motor. For example, if the voltage is reduced 10 per cent on a motor, the torque it will develop will be only 81 per cent of that at normal voltage. It is therefore readily seen that if the voltage is materially reduced, the motor may not be able to start its full load, or if it does start the load, acceleration will be slow. When the elevator makes floor-to-floor stops, the greater part of the time acceleration is required. Therefore anything that lengthens the acceleration period will slow up the service.

Overvoltage on an a-c elevator motor may also have objectionable effects. A 10 per cent increase in voltage above normal causes the motor to develop a torque 121 per cent of normal, which may increase the acceleration rate of the elevator to the point at which it will be unpleasant for the passengers in the car. It may also cause excessive noise in both the motor and controller. High voltage increases the current through the magnet coils and increases the pull on the contactors, which may cause them to slam unduly, resulting in increased maintenance needs and costs.

Checking the Power-circuit Voltage. When checking voltage on an elevator, it is not sufficient to take this reading at any particular load or time. Obtain a record for a period of a day or a week on a recording voltmeter. It is better still if one of these meters can be connected to each phase for the test period. It is quite as important

FIG. 15-2. Drum-type elevator machine driven by wound-rotor motor.

that the voltage be balanced on all phases as that it be correct on any particular phase. Where power is supplied over long transmission lines, unless automatic regulation is provided there may be wide variation in the voltage over the 24 hours of the day. In the daytime, when the load is heavy on the system, the voltage may be low, and at night or during other light-load periods the voltage may be high. A variation in the voltage, applied to an a-c motor, of from 10 per cent above normal to 10 per cent below normal will cause the torque to be 50 per cent higher at the high voltage than at the low voltage; that is, 121 per cent torque is 50 per cent greater than 81 per cent.

With this wide variation in torque it may be impossible to adjust the control equipment to give satisfactory operation at all periods of the day. Adjustments that would give smooth acceleration at the low voltage may cause the motor to start with a jerk at the high voltage. On the other hand, adjusting the control for satisfactory operation at the high voltage may result in slow acceleration at low voltage, or the motor may actually stall.

Effects of Frequency Variations. If trouble is experienced with the torque and speed of an a-c motor, the frequency of the power system should also be checked. The speed of these motors varies directly as the frequency. A change in frequency also affects the

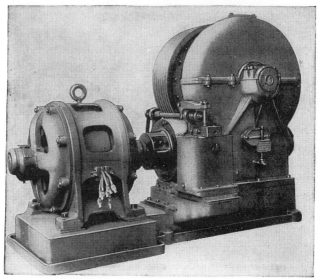

Fig. 15-3. Geared traction machine driven by two-speed squirrel-cage motor.

magnets on the controller. An increase in frequency will decrease the current through the coils and vice versa for a decrease in frequency. With a decrease in frequency, current increase in the magnet coils may cause excessive heating. From the foregoing it is evident that close voltage and frequency regulation are desirable for a-c elevator motors and controllers. On modern power systems, voltage and frequency are usually closely regulated, so that little trouble develops from these sources.

Noisy Operation. Noisy operation of a-c motors is more prevalent than of d-c motors. Therefore, anything that might occur in a-c motors to cause unusual noise may be doubly objectionable. In any motor, if the rotating member is out of balance, it will cause vibration and noise in operation. With an a-c motor, in addition to mechanical vibration there is also magnetic vibration, and the latter is more likely to cause trouble than the former. Mechanical vibration, once eliminated, is not likely to occur again, but magnetic vibration may be produced from a number of causes that can develop in the normal operation of the machine.

Before an attempt is made to diagnose a case of noisy operation, it is well to determine if the cause is mechanical or magnetic. This may be done by disconnecting the motor from its load and then connecting it to the line. After it comes up to speed and is operating unusually noisily, if the switch is opened and the motor continues to vibrate, it will be known that mechanical causes are producing the noise. On the other hand, if when the switch is opened the noise immediately disappears, current in the windings is the cause of the trouble.

Worn bearings or a loose rotor on the shaft may cause noisy operation of the motor. When bearings become worn and allow the rotor to get too close to the stator core, the motor may fail to start and develop heavy vibration. If the motor remains connected to the line under this condition, vibration may injure the windings, by excessive heating, by the vibration itself, or both. With a squirrel-cage rotor, loose connections between some of the bars and end rings will cause unequal distribution of the current in the rotor and may produce noisy operation. Vibration from this cause generally disappears when the motor comes up to speed. In a wound-rotor motor that had a number of grounds in the rotor winding, noisy operation developed.

Other causes of noisy operation are: voltage too high; voltage unbalanced; open circuits in part of the stator windings; short circuits in part of the stator windings, generally accompanied by excessive heating; and single-phase operation. The last is generally due to causes outside the motor, such as an open phase in the external wiring.

Failure of the Motor to Start. Failure of an a-c motor to start when the load is free to move may be caused by interruption of the power supply due to an opening of the protective devices or to an opening in one phase of the wiring. If the power supply is completely cut off from the motor, when the starting switch is closed a humming noise will not be emitted by the motor. If one phase is open, the motor will not start but will develop a loud roaring noise. Worn bearings that allow the rotor to come in contact with the stator will prevent the motor from starting; this fault will be accompanied by a loud roaring noise when power is applied to the stator. In this case, sometimes the motor may start, and sometimes it may not.

If a wound-rotor motor is connected to the line with the rotor resistance cut out, the motor is very likely not to start, but will take a current from the line that will open the protective device. Opens in the resistance of a wound-rotor motor will cause it to fail to start, but this condition will be accompanied by a loud roaring noise. A similar condition would exist with a squirrel-cage motor if a number of the bars were broken from the end rings. Grounds in the stator windings or short circuits between phases may cause failure to start, although the conditions will generally cause the protective device to open.

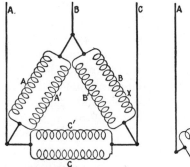

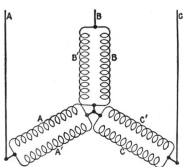

Fig. 15-4. Diagram of two-parallel delta-connected winding.

Fig. 15-5. Diagram of two-parallel star-connected winding.

Low Torque. Low torque may be caused by low voltage, poor connection between rotor bars and end rings of squirrel-cage motors, open circuits, short circuits, or grounds in the rotor windings of wound-rotor motors. Open circuits, short circuits, or grounds in the stator windings may cause low torque in addition to other effects already referred to. If one phase is completely open, then the motor will not start. Many motors are wound with multiple-circuit windings. For example, assume that each phase has two windings in parallel, as shown in Figs. 15-4 and 15-5. If one of the windings is open at X (Fig. 15-4), there are two groups of windings in parallel between leads A and B and between leads A and C, but only one winding between leads B and C. Such a condition in the windings reduces the torque and causes the motor to take an unbalanced current from the line. Ammeters connected in the circuit would show the current to be unbalanced, but this should not be taken as an indication of a fault in the windings, since unbalanced line voltage would have a similar effect.   If the motor is started cold and allowed to run for a short time, the phase with only one winding in circuit will heat more than the others.   However, after it has been decided that the cause of the trouble is not external, the surest test is to break the windings into their parallel groups and test through them with a lamp.

Testing for Faults in Motor Windings.   If low-voltage direct current is available, the windings may be excited from this source and readings taken on coil terminals with a millivoltmeter.   If a section of winding is open-circuited, no reading will be obtained on the coils until the open is bridged by the meter, when a heavy deflection of the needle will occur.   In making this test take readings on coils only and not between phases, or misleading results may be obtained.

With a delta-connected winding, the whole winding is excited by connecting any two leads to the power source, where with a star-connected winding (Fig. 15-5) only two phases are excited.   In the delta connection one phase will carry about double the current of the other two; consequently the readings obtained on all three windings will not be equal.   The winding carrying the largest current will give the greatest deflection on the millivoltmeter.   To connect the voltmeter to the coil terminals, sharp steel points can be used as test lead terminals to pierce the insulation and make contact with the coils without removing the insulation.

Causes of Low Speed.   Low speed may be caused by low frequency, low voltage, rotor rubbing on the stator, poor connections or open circuits in the rotor windings, or open circuits or short circuits in the stator windings.   The latter will generally be accompanied by excessive heating of the windings.

Heating of the windings may also be caused by a high voltage open circuit in part of one phase, as previously explained, and the rotor rubbing on the stator.   In elevator service overloads will rarely be the cause of heating.

With a-c motors, if a phase is reversed accidentally on the power system, it will cause motors to reverse.   This is a very dangerous condition on elevators and should be protected against with reverse-phase relays.   Such relays will generally also protect against single-phase operation and should be installed on every polyphase a-c motor used on elevator service.   Reversal of phases on a-c power systems has been the cause of wrecking elevator machinery.

In case of a reversed phase outside the motor room, the fault can be corrected by crossing the leads of any one phase at the power switch.   However, do not do this until it is known how the phase was reversed on the line and what is being done to correct the trouble. If the trouble is corrected at the motor's switch and the elevator put into service, and then to make conditions normal on the line a phase is reversed, the elevator will again be operating in the wrong direction on the controller. To avoid such an occurrence, what is going to be done on the line should definitely be determined.

Speed Too High. It is quite general practice, where single-speed high-resistance squirrel-cage induction motors are used on slow-speed elevators, to start the motor with a resistance connected into the primary circuit, which is cut out of circuit after a given period. This is generally cut out in one or two steps by contactors. Closing of these contactors is usually controlled by timing relays. It is essential for these relays to be kept in operating condition, for if the stator resistance is not cut when a heavy load is lifted, the motor may not be able to start the load, and when a heavy load is lowered, the elevator may race. When a heavy load is lowered, the induction motor becomes a generator, pumps back into the line just as a d-c motor does, and acts as a brake to keep the elevator under control. If the primary resistance is in circuit, the motor's speed will have to increase considerably above normal to supply the necessary braking current.

An open circuit in one leg of this primary resistance will prevent the motor from starting. Then, when the resistance contactor closes, the motor will start, but it will probably cause an unpleasant movement of the car. Where a full-magnet type of controller is used, the direction switches are magnetic contactors controlled from the car switch or push buttons, as on d-c motor controllers. What has been said in Chap. 14 regarding the troubles in the direction-switch operating circuits for direct current applies to a-c controllers.

High-speed Equipment. When two-speed motors are used they are generally of the squirrel-cage type. Usually all starting of these motors is done on the high-speed winding, and the slow-speed machine is switched in during stopping to slow the machine down before disconnecting from the line and applying the brake to stop the machine. Except for the fact that control is provided for two motors instead of one, and that sometimes a governor-operated switch holds the slow-speed motor's contactor in after the high-speed motor has been cut out, until the elevator has slowed down, the control for these motors does not offer difficulties not found with other types of full-magnet car-switch or push-button controllers. With these equipments, if trouble is experienced with the brake running hot or not stopping the car quickly enough, make sure that the slow-speed motor cuts into service during stopping. If this is not done, all the work of stopping the elevator from high speed will have to be done on the mechanical brake, which may result in unsatisfactory operation.

Faults in Contactor Magnets. Alternating-current magnets require careful adjustment if they are to operate quietly. They are usually designed with a closed magnetic circuit. Pole faces and armature are ground to a good fit. Any dust, lint, or oil collecting on these surfaces that prevent the armature from seating properly may cause the contactor to be noisy. In the pole faces of these contactors there is usually a copper loop, known as a shading coil. If this loop breaks it will cause noisy contactor operation. Some controls use rotating magnets (Fig. 1-47) which function like small polyphase induction motors and which may be called nonsealing magnets.

Current taken by a coil on an a-c circuit is limited by the ohmic resistance of the coil's circuit and by the counterelectromotive force induced in the coil, just as in the primary windings of a transformer. When the coil circuit first closes, the contactor is open; consequently there is a long air gap in magnetic circuit. This reduces the magnetic field in the core, which reduces the counterelectromotive force in the coil so that the current flowing will be larger than when the magnetic circuit is completely iron, as when the contactor is closed. On this account, if the current is left on a contactor coil, when the contactor does not close it is likely to burn out in a short time. If trouble is experienced with a contactor coil burning out, check the possibility of the contactor not closing. On direct current, a coil might have a small part of its winding short-circuited without causing serious trouble. On alternating current the short-circuited section of a coil becomes a short-circuited secondary of a transformer and might prevent the coil from closing the contactor or, if the contactor does close, the current flowing might be sufficient to cause the coil to burn out in a short time.

CHAPTER 16

Lubrication

General Considerations. The problem of elevator lubrication will vary with the type of machine and the conditions under which the equipment operates. On drum and geared traction machines, lubrication of the worm gear is important in their successful operation, whereas with the direct-traction type the worm gear does not come into the picture. Broadly, lubrication of elevator equipment may be divided into ropes, worm gear, guide rails, sheaves, motor and other bearings, and wearing parts.

Lubrication of Ropes. A wire rope is a complex mechanism, composed of a number of wires twisted to form a strand, with a number of these strands then twisted about a hemp center to form the rope. The combinations of numbers of wires per strand and numbers of strands in a rope are practically unlimited. There are on the market about seventy of these combinations, but for elevator service 6 strands of 19 wires each or 8 strands of 19 wires each are the types generally used.

In elevator service the ropes bend over sheaves, and while the car is in motion the ropes bend and straighten continuously. This operation causes a continuous slight rubbing of the wires and strands upon each other. These rubbing surfaces, like any other, require lubrication. When the rope is made, the hemp center is thoroughly saturated with lubrication and, if it is not installed in a damp place, this will furnish sufficient lubrication for a considerable period.

There is a difference of opinion about the application of external lubrication, but it has been proved by tests that internal friction in ropes increases the bending stresses that cause wires to break. This fact alone should be sufficient to warrant lubrication of elevator ropes. Where wires on the outside break, they can be detected before the rope gets in a dangerous condition. However, when the inside wires break, as they sometimes do owing to lack of lubrication, ropes may reach the danger point without giving external evidence. Ropes operating in vapor, fumes, or damp places will rust and corrode unless they are kept lubricated.

There seems to be little reason for not lubricating elevator ropes when the advantages are weighed against the objections. About the most serious objection is that of the extra work and the care that must be exercised to prevent the lubricant from dropping onto the car. A good nonacid medium-heavy oil that will penetrate to the core and also stick to the surface is desirable. There is not much value in a lubricant that is so thin that it will run easily and drip

FIG. 16-1. Lubricating a traction-elevator hoist rope with a brush.

from the ropes. On the other hand, it should not be so thick and sticky that it will not penetrate into the hemp center. A number of elevator manufacturers have lubricants for sale that they recommend for this purpose. Experiments have proved that a nonacid lubricant having a Saybolt viscosity of from 1,000 to 2,000 sec at 210° is suited for this purpose.

Some authorities claim that when the ropes run in a warm, dry place for less than three years over generous-sized drums and sheaves, there is sufficient lubrication in them to prevent abrasion of the wires and strands on each other. If the life of the ropes exceeds this period or if they travel over small sheaves or drums, more frequent lubrication is necessary, particularly in cold, damp places. The best practice is to watch the ropes carefully and to apply lubrication if they show signs of being dry or rusting. Even in dry places it will be found good practice to apply lubrication at least once every six months; every three months is preferable under more severe conditions. There are several ways to apply lubricant to the ropes. One is to paint it on with a brush, as in Fig. 16-1. However, do not overlubricate, as this may cause rope slippage on the traction sheave and oil dripping on the car.

Fig. 16-2. Section through worm and gear showing gear-case oil level and ball thrust bearing at *A*.

Worm and Gears. Worm gears on drum machines and on geared traction machines require lubrication. In this service there are rubbing surfaces under heavy pressure that are in contact at one moment and are relieved of the load at the next. In some applications ball or roller guide and thrust bearings are used on the worm shaft, as at *A* in Fig. 16-2, whereas in others, sleeve guide bearings and disk thrust bearings are employed. In still other applications, a combination of the two types of bearings may be found.

Oil used in the gear case must have sufficient body to stand the heavy pressure to which it is subjected on the gear teeth without allowing metal-to-metal contact with the worm. It should also be free from alkalies and acids. This is particularly true where ball or roller bearings are used, since their service efficiency is seriously impaired by even slight pitting of their surfaces. Elevators are frequently located where they are subjected to wide ranges of temperature, from the intense heat of summer to the lowest temperature of winter. In such cases it is important for the oil not to be seriously affected by a change in temperature.

Successful lubrication of worm gears has been accomplished with a straight mineral oil, such as a high-grade steam-cylinder stock having a flash point of about 600°F and a viscosity of about 150 sec Saybolt at 210°F. Other recommendations are for a straight mineral oil having a viscosity of about 120 sec at 210°, but a pour test much lower than cylinder stock. Oil containing such fillers as graphite has been used. Then again, castor oil has many advocates, and as high as 50 per cent castor has been used in worm-gear lubricants. Where castor oil is mixed with a straight mineral oil, there is always the danger that they will separate in the gear case. This can be prevented if the oils are properly compounded, but it will probably require the use of a third component. In view of the wide variation of opinion it would be well to consult the elevator manufacturer or a responsible lubrication company's engineers if trouble is experienced with the problem. Most elevator manufacturers have gear lubricants that they recommend for their machines.

There is an idea among many maintenance men that the higher the oil level is in the gear case without running out, the better the worm and gear is lubricated; this is incorrect. Under these conditions the gear acts as a pump and not only increases the machine's power consumption, but also heats the oil, and it works out of the gear case to create a dirty mess around the machine. For best lubrication of the worm and gear, oil level should not come above the worm-shaft center line. Some manufacturers arrange their machines so that only the lower side of the worm dips into the oil (Fig. 16-2); this is a section through a gear on machines operating at car speeds of up to 500 fpm. The worm in this machine has seven threads. To avoid putting too much oil into the gear case, a large overflow is provided, as shown at *A* in Fig. 16-3. This overflow also provides a means of determining the amount of oil in the gear case.

Preventing Cutting of Worm and Gears. Owing to the lack of lubricant or the use of an improper quality, a worm and gear may start to cut. This action is evident from a loud grinding noise in the gear case and heating of the worm and gear, accompanied in many cases with vibration in the machine. Do not confuse the grinding noise with that from a ball thrust bearing with chipped

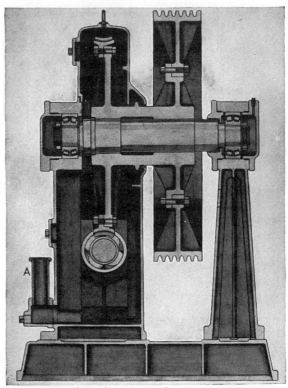

FIG. 16-3.   Cross section through V-grooved gear-type traction machine showing large oil overflow *A* on gear case.

balls or races. To stop a worm and gear from cutting, put about 1 lb of sulfur in the gear case with the oil and operate the machine until the gear teeth take a smooth polished surface. Under ordinary conditions this will require about eight hours of operation. If the elevator is used only intermittently, a longer period of time will be required. After the worm and gear are in condition, drain the oil and wash out the gear case with kerosene, being sure that it is thoroughly cleaned out before putting in new oil.

Motor Bearings and Other Parts. For the motor bearings and other parts, except for slow-moving sheaves equipped with mill-type babbitt bearings, a good grade of motor oil will generally meet the requirements. On slow-moving sheaves carrying heavy loads, a heavy oil or grease is generally used, depending upon the method of applying the lubricant.

When the elevator equipment is lubricated, the safeties under the car should not be overlooked. These parts are seldom called upon to operate in normal operation, but since when they are called upon it is generally in an emergency, they should be kept in good working condition and not allowed to rust and become inoperative through lack of attention and lubrication.

When purchasing and applying lubricants, remember that a good oil properly applied is the cheapest kind of maintenance. Buying a lubricant because it is cheap is false economy.

Lubrication of Guide Rails. Lubrication of guide rails is the most difficult problem connected with passenger- and freight-elevator operation. The lubricated surfaces of the rails are exposed and collect dust and dirt blown about by the car as it passes up and down the hoistway.

If the oil film becomes too thin or broken, the guide shoes bite the rails and cause noise and unpleasant riding in the car. Wear of the guide shoes, power consumption for the elevator's operation, and maintenance costs will increase. When there is an excess of lubricant on the rails, it is thrown off by the guide shoes onto the car and into the bottom of the hoistway. If oil gets on the car, it is likely to work itself inside of the cab and cause stains or, which is worse, get on passengers' clothing. In addition it creates a dirty condition in the hoistway, adds to the difficulties of keeping the place clean, and creates a serious fire hazard.

Manual Lubrication. The old method of lubricating guide rails is to paint them with a heavy oil or grease periodically, the operation being performed by an attendant riding on top of the car. Lubrication of guide rails by hand is about the most hazardous and disagreeable job in the maintenance of elevators. Not only are two men required to do the job, one to operate the car and another to ride on top and do the greasing, but the elevator must be taken out of service while the work is done. For these reasons, where lubricating is done by hand, it is frequently neglected.

Another objection to hand greasing is that the rails either have a feast or a famine of lubricant. There is generally too much grease on the rails when lubricating is completed. After the elevator has operated awhile there is a deficiency of lubrication, and the rails may not be greased again until attention is called to the need by the guide shoes biting the rails.

Automatic Guide-rail Lubrication. In an attempt to overcome these difficulties, automatic guide-rail lubricators may be used. Automatic lubrication of guide rails is by no means a simple problem, as evidenced by the number of devices that have been developed for this work. Lubricants ranging in consistency from light oils to solid greases are used in automatic lubricators. One of the advantages claimed for light oils is that they tend to run down the rail and wash the dust and dirt off. On the other hand, at high speeds they are more likely to be thrown off the rails by the guide shoes. This difficulty can be at least partly overcome by chamfering the top and bottom edges of the guide-shoe gibs, so as to form receptacles for the oil. Greases, although they adhere to the rails better than oils, have the disadvantage of holding any dirt that may be deposited on the rails.

Higher car speeds and rises have made guide-rail lubrication more difficult. Automatic lubricators satisfactory for medium-rise, medium-speed cars may not do so well in faster, higher-rise service. Passengers have also become noise-conscious and cars classed as fairly quiet a few years ago may now be rated noisy and unsatisfactory.

Guide-shoe Gibs. These are an important factor in guide-rail lubrication and smooth car riding. Early guide shoes were made of cast iron, unlined and bolted solidly to the car crosshead. Later they were babbitt-lined, either cast in place or made as a separate replaceable lining or gib. These guide shoes were made adjustable and held to the rail by spring tension, a standard method used today. But many other materials are used now for gibs, varying with operating conditions, such as cast iron, oil-impregnated wood, wood with babbitt inserts, horn fiber, horn fiber with lubricating compound inserts, various plastics including nylon fiber or wood faced with bronze, nickel-copper alloys, and other metals.

Impregnated fibrous materials or wood gibs give smooth riding and better guide-rail lubrication. Oil-impregnated maple gibs have these qualities in high degree, but lack wearing ability and may split under severe operating conditions. Holes filled with babbitt on their inner surfaces improve their wearing qualities.

These gibs, like most nonmetallic designs, are made in three parts, two sides being doweled to a back. Horn-fiber gibs are similar in design to wood gibs but don't need babbitt inserts because the wearing qualities of fiber are about equal to those of ordinary babbitt.

Other gibs are made of a tough elastic molded product with a canvas base and have face inserts of a lubricating compound. This material is not easily broken, chipped, or cracked in service. Rubbing on the rails gives these gibs a highly glazed surface. For slow- and medium-speed cars operating on good rails, they are practically self-lubricating.

This design has been used in several large buildings in which guide rails are lubricated only at long intervals. Further, the holes on the rubbing surfaces tend to pick up grease where there is an excess and redistribute it over the drier parts of the rail. Notches cut in the gibs' ends, into which grease can be packed, help lubrication. With high-speed cars, however, guide rails need additional lubrication when these gibs are used in car guide shoes, just as with other designs, except that nonmetallic gibs are less likely than metal to bite a dry guide rail. This simplifies the lubrication problem.

Guide-rail Conditions. Before any gib will run right, the guide rails must be in line and their surfaces and joints smooth. Rail running surfaces can be smoothed by having cast-iron gibs in the guide shoes and the rails lubricated with a mixture of sulfur and oil. Run the car until the rails have a good surface. Then clean them thoroughly with kerosene. Replace the guide-shoe gibs, lubricate the rails with regular lubricant, and start normal car operation.

When the car does not hang in good balance between the guide rails, parts of the guide-shoe gibs wear more rapidly than they should. Several ways of attaching the hoisting ropes to the car, particularly with automatic rope equalizers, provide for shifting car balances so that the car hangs true.

Operation without Lubrication. Guide-rail lubrication has always been a troublesome elevator maintenance problem. A lubricant is difficult to apply satisfactorily under all conditions. It creates a fire hazard in the hoistway, and the hoistway walls and outside of the car are difficult to keep clean. This has led to the development of guide-shoe gibs and guides that can operate dry.

One of these, a laminated fiber gib (Fig. 16-4), consists of a stack of U-shaped laminations, each about $\frac{1}{8}$-in. thick. These are assembled with a graphite grease compound between them on four rods. The U fits on the guide rail and the graphite supplies limited lubrication. Graphite guide-shoe gibs have also been used.

Rubber-tired Rollers. Figure 16-5 shows one design of roller guide. It has three large-diameter rubber-tired rollers *R*, one on each face of the rail. Each roller runs on a ball bearing supported by a rocker arm *A*, pivoted at *P*.

Continual contact between rollers and rail is maintained by springs *S* on each rocker arm. This prevents the rubber tires from scuffing and developing flat spots. Each tire is so shaped that it cannot overlap the rail faces or interfere with the tires on other rollers.

Another design of roller guide has rollers made of canvas bonded with rubber to form a tough flexible tread. Each roller runs on a prelubricated ball bearing, clamped between two side plates. Face and side roller assemblies have eccentric spindles for adjusting roller

FIG. 16-4.   Part of laminated fiber guide-shoe gib.

pressure against the rail. Rubber bushings around the spindles reduce vibration and noise.

Other roller guides have six ball-bearing rollers, two on each rail face, mounted at the end of three rocker arms. These arms are flexibly mounted at their centers with springs to give them a knee action so that the rollers can adjust to any roughness or misalignment of the rail.

Roller guides may be used on both car and counterweights, and are installed on most new elevators. On the car, the rollers of one design are 6 in. in diameter for speeds up to 500 fpm, and 10½ in. in diameter for higher speeds. Counterweight rollers are about one-half the size of the car-guide rollers. Other sizes of rollers are also used, down to 3 in. in diameter.

For successful operation of roller guides the rails must be free of oil and grease. Before installing roller guides on rails that have been lubricated, clean the rails thoroughly. Wash them first with kerosene, then with carbon tetrachloride.

Operating on dry rails and using a fire-resistance traveling cable make the hoistway equipment practically fireproof. Reports on

Fig. 16-5.   Rubber-tired roller guide operates on dry guide rails.

hoistway fires where rails are lubricated indicate they are more serious than generally believed, and anything that can be done to prevent them is a good investment. Power reduction with roller guides varies with conditions, but in some installations has been reported to be as high as 40 per cent. There is also a saving in lubrication and maintenance cost in the hoistway.

CHAPTER 17

Ropes, Their Construction, Inspection, and Care[1]

Metals Used in Wire Ropes. Wire rope is an important part of electric-elevator equipment. It connects the car and counterweights with the hoisting machine and overspeed governor to the car safety devices. Therefore the car's operation and the passengers' safety depend upon wire rope. For this reason those responsible for elevator operation should be familiar with wire-rope constructions and factors affecting its service and safety.

Wire rope is made in a great variety of forms and of many materials. Although generally made of some grade of iron or steel, Monel metal, bronze, and other metals are used for special conditions. For example, Monel-metal rope is a noncorrosive type, suited for wet places, such as meat-packing houses or places where chemical fumes or salt in the atmosphere would destroy iron or steel rope.

The material used in so-called iron rope is a very mild steel, containing about 0.1 per cent carbon. It is comparatively soft, ductile, and of low tensile strength. Wire of this material for rope construction has a tensile strength of about 85,000 psi.

What is known as traction steel, a form of toughened mild steel containing about 0.35 per cent carbon, is used extensively for traction-type elevator ropes. This material has a tensile strength in the wire of about 150,000 to 170,000 psi. Cast steel, the first of the so-called higher-carbon steels, frequently called crucible steel, is another common material used in wire-rope construction. It is one of the most ductile of the high-carbon steels, and has a tensile strength in the wire of from 170,000 to 220,000 psi. Three other steels,

[1] For assistance in the preparation of this chapter the author is indebted to the American Cable Company, the American Wire & Steel Company, John A. Roebling's Sons Company, Hazard Wire Rope Company, A. Leschen & Sons Rope Company, Williamsport Wire Rope Company, Broderick & Bascom Rope Company, MacWhyte Wire Rope Company, and the Bethlehem Steel Company.

designated as mild-plow, plow, and improved-plow are in common use for wire-rope construction. They have a tensile strength in the wire of about 220,000 to 280,000 psi. These and others, appearing under various trade names, are mostly some form of open-hearth steel.

Of the various materials for wire-rope construction, iron and mild steel are about the only ones used for elevator ropes. Mild-steel ropes frequently appear under the trade name of traction steel or other designations.

Rope Constructions. After the wire has been drawn to size and heat-treated to give it the desired qualities, it is ready to be formed into rope. Rope-making consists of twisting a given number of wires into a strand and laying a number of these strands about a hemp center. In some cases, for use on dead loads, disk conveyors, and hot-metal cranes, a wire center is used. Tiller rope and mooring line have strands made with hemp centers. In general the

Fɪɢ. 17-1.   Regular 6 x 19 wire-rope construction.

strands are laid around a central wire, although in some constructions they have centers of special forms.

Combinations of wires, their sizes in a strand, and the number of strands in a rope are almost unlimited. There are on the market about 80 different constructions varying from 3 to 91 wires in a strand and from 3 to 19 strands in the completed rope. Only a few of these constructions are in general use for elevator service.

A common hoisting-rope construction is one that has 6 strands of 19 wires per strand, known as the 6×19 regular construction (Fig. 17-1). Each strand has a central wire about which is placed a layer of 6 wires. Outside the 6-wire layer is another of 12 wires, both layers being twisted in the same direction. The rope is formed by laying 6 of these strands about a hemp center, as in Fig. 17-1.

Another common construction has 6 strands of 37 wires each. Where the wires in the strands are all approximately the same size, the construction is the same as for the 19-wire strand, but with a third layer of 18 wires. This is the highest number of wires per strand used for elevator-hoisting ropes.

For certain purposes where a very flexible rope is required, 61 and 91 wires per strand have been used. The 61-wire strand is obtained by adding a layer of 24 wires over a 37-wire construction. Adding a 30-wire layer over the 61-wire construction produces the 91-wire strand. These various constructions are shown in Fig. 17-2. The greater the number of wires per strand for a given size rope, the smaller the individual wires will be in cross section.

Except for hawsers, mooring lines, and guys, few wire ropes are made with all wires the same size. Even when the wires may appear to be the same size, as in the 6 × 19 regular constructions, the inner ones are slightly larger than the outside ones. This difference in size is necessary in a rope of this kind, to obtain a construction that will stand bending and retain the original arrangement of the wires in the strands. Another rope used for elevator service has 8

Fig. 17-2.   Strand construction, ranging from 7 to 91 wires.

strands of 19 wires each. Although this construction is more flexible than the 6 × 19, it has the same strength.

Warrington Construction.   Figure 17-3 shows the cross sections of various rope constructions used in elevator service. A and B are the regular 6 × 19 and 8 × 19 types, respectively. All wires in either of these are approximately the same size. Constructions C and D have three sizes of wires. The 7 inside wires are approximately the same diameter and are surrounded by 12 that are alternately large and small, a construction known as Warrington. Experience has shown C and D to be satisfactory constructions, and they are used extensively.

Where bending stresses are severe, the 6 × 37 construction, shown at E, is sometimes used. Each strand is made of 37 wires of approximately the same size, and 6 strands are laid about a hemp center to form a rope. Since the wires are small, the rope is not suited to conditions where wear is an important factor in its life.

Seale Construction.   Another construction, used for elevator-hoisting ropes, is known as Seale; three different combinations, F, G, and H, are shown in Fig. 17-3. The 6 × 19 Seale construction has 1 large center wire surrounded by 9 small ones, around which are 9 large wires. This construction will, for many conditions, stand more wear than constructions *A* and *C*. The rope is somewhat stiffer than the other constructions and is not suited for use on small sheaves or for places where it would be subjected to short reverse bends.

The 6 × 27 Seale construction *G* (Fig. 17-3) has 7 small wires in the center of each strand, with 10 small wires and 10 larger ones laid over the 7 center wires. This is a more flexible construction than that obtained with the 6 × 19 Seale.

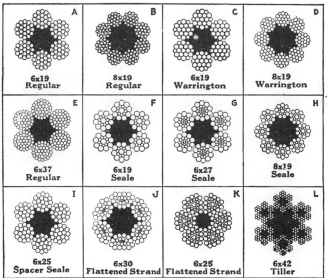

Fig. 17-3. Cross-sectional arrangements of different types of wire ropes.

At *H* is shown an 8 × 19 Seale construction. The strands in this rope are made the same as those for the 6 × 19, but with smaller wires. The 8 × 19 construction is more flexible than those shown at *F* and *G*. The Seale construction is such that the crowns of inside wires fit into the valleys between the outside wires. This allows a close construction of the strands, resulting in an increased metallic cross-sectional area and strength. The 8 × 19 rope has superior bending qualities compared to those of the 6 × 19 construction. Where ropes are subjected to reverse bends, the 8 × 19 is, in many cases, preferable to the 6 × 19 construction.

Another Seale construction is known as spacer Seale, a 6 × 25 section being shown in Fig. 17-3 at *I*. This construction has 6 wires twisted around a center wire, all wires being approximately the same size. Six small spacer wires are placed in the valleys between the 6 large wires laid around the center. Over these are laid 12 wires of approximately the same size as the center ones. The 6 small spacer wires act as keys to maintain a fixed relation between the outside and inside wires in the strand.

Flattened-strand Construction. A flattened-strand construction is sometimes used to obtain a rope with a large wearing surface. Two types of these are shown in Fig. 17-3 at *J* and *K*. The center of the strand, construction *J*, is made of 9 small wires arranged to form an ellipse. About this center are laid 9 larger wires, over which is placed a layer of 12 wires. These wires are slightly larger in diameter than those used in the regular 6 × 19 rope construction. This rope has a larger wearing surface than ordinary types and is more flexible than the regular 6 × 19 construction.

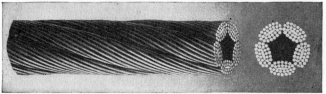

Fɪɢ. 17-4.   Flattened-strand rope construction with elliptical-ribbon center.

Another flattened-strand construction, somewhat similar to *J*, has an elliptical-ribbon center (Fig. 17-4), but is not as flexible.

Triangular-strand Construction. In the construction shown in Fig. 17-3 at *K* the strands are made with a triangular center wire about which are twisted 12 small wires. Over these are laid 12 larger wires. This gives a strand having a form approximating an equilateral triangle. Six of these strands are laid about the usual hemp center to form a rope. These flattened-strand constructions have about 50 per cent more wearing surface than the round-strand rope, and *K* is stronger than round-strand types of the same materials.

Tiller Rope. Where a very flexible type is required, such as for the operating rope on hand-operated elevators, what is known as tiller rope is used, as shown in Fig. 17-3 at *L*. This has 6 strands, each constructed like a 6 × 7 hemp-center rope. These strands are laid about a hemp center, making a very flexible arrangement, but the rope's strength is only about one-half that of the other constructions of the same diameter.

Regular and Lang Lay. Numbers and sizes of wires, their arrangement in the strands, and the number of strands tell only a part of the story of a rope's construction. The way the strands are laid to form the rope must also be known. There are two lays, regular and Lang. These lays may be either left or right. A right-lay regular-lay rope is shown in Fig. 17-5 at *A.* In this construction the strands are laid to give the appearance of a long-pitch right-hand thread on a bolt. A nut threaded to fit the pitch of the lay could be threaded onto the rope by turning it in the clockwise, or right-hand, direction. What has been said regarding a right-lay regular-lay rope applies to a left-lay regular-lay, shown in Fig. 17-5 at *B,* except that everything is left-handed. Lang-lay rope can be obtained in either right or left lay, as shown in Fig. 17-5 at *C* and *D.*

Fig. 17-5. Wire-rope lays in general use. *A,* regular-lay right-lay. *B,* regular-lay left-lay. *C,* Lang-lay right-lay. *D,* Lang-lay left-lay.

Regular-lay rope has its strands twisted in an opposite direction to the rope lay. For example, in the right regular-lay rope (Fig. 17-5, *A*), the strands are twisted in a clockwise direction, but the rope has a counterclockwise twist. In a Lang lay the strand and rope twist are the same, as shown at *C* and *D* in Fig. 17-5. A right Lang lay *C* has the strands and the rope twisted in a counterclockwise direction. In a regular-lay rope the outside wires on the strands run almost parallel with the axis of the rope, but in a Lang lay these wires tend to follow around the rope with the strand. These characteristics are clearly seen by comparing *A* and *B* with *C* and *D* in Fig. 17-5.

Lang-lay rope is more difficult to handle than regular lay, as it has a greater tendency to unstrand. It is also more easily kinked and bird-caged than regular-lay rope. For these reasons Lang-lay rope has not been used extensively on elevators. It is, however, more flexible than regular-lay rope, and the outside wires on the strands present a greater wearing surface than those of regular-lay rope, as can be seen by comparing Figs. 17-6 and 17-7. Most flattened-strand ropes are Lang lay, such as shown in Fig. 17-4. A worn flattened-strand rope is shown in Fig. 17-8, which indicates its large wearing surface.

Preformed Rope. Wire rope, as ordinarily constructed, has a tendency to unlay unless it is seized at the ends to prevent the strands from untwisting. There are on the market wire-rope designs in

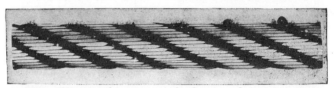

Fig. 17-6. Partly worn regular-lay round-strand rope.

Fig. 17-7. Partly worn Lang-lay round-strand rope.

Fig. 17-8. Partly worn flattened-strand rope.

which the wires are preformed to fit the twist in the strands and the strands are preformed to fit the twist in the rope. These are sold under various trade names for use on elevators, but are generally known as preformed rope. This type of rope does not have a tendency to unlay. In fact, a strand may be taken out of a section and a wire taken from the strand. These may be returned to their original positions in the rope without disturbing the construction. These brands should not be confused with the nonrotating types.

A right-lay regular-lay construction is the one commonly used. In some cases left-lay ropes are used for oil-well drilling. For some elevator and hoist applications, left-lay is used in combination with right-lay rope, to produce a nonrotating combination.

Nonrotating Ropes. As wire rope is commonly constructed, it will untwist if allowed to support a free load; in other words, it will rotate the load. There are nonrotating ropes made, but these are special constructions, one of which is shown in Fig. 17-9. This rope contains 18 strands of 7 wires each. The center of the rope has 6 strands left-lay Lang-lay about a hemp center. Around the center strands are 12 strands placed right-lay regular-lay. From this it will be seen that the two layers tend to twist in opposite directions and prevent rotation.

Figure 17-10 shows a rope construction, the strands of which are twisted alternately left and right. Thus, when the strands are laid

Fig. 17-9.   Two-layer combination Lang- and regular-lay nonrotating rope.

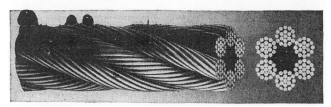

Fig. 17-10.   Single-layer combination Lang- and regular-lay nonrotating rope.

into a rope, they give a combination Lang and regular lay. When the rope is loaded, one lay tends to twist in one direction, while the other has the opposite effect and prevents rotation. There is a flat-strand-rope construction, shown at *J* in Fig. 17-3, that is made nonrotating by the use of alternate strands twisted in opposite directions, giving a combination regular and Lang lay. The chief use for these ropes is on conveying machinery, but they are mentioned here as interesting constructions. In the foregoing only a few of the more common types of rope have been considered, particularly those that have been used for elevators.

Measuring Size of Ropes. If the diameter of a rope is not known, it will have to be measured. Some wire ropes are made so that their external surface is almost a true circle, in which case they have practically one diameter. For the ropes generally used on elevators this is not true, and there is considerable difference between their maximum and minimum diameters. The maximum diameter (Fig. 17-11) is the correct one to use when specifying wire rope. In other words, the diameter of a rope is the diameter of the minimum diameter circle through which it can be passed.

A used rope will have a smaller diameter than it had when new, due to the bedding of the strands into the hemp center, to stretching, and to wearing of the strands. Sizes of hoisting ropes below $1\frac{1}{16}$-in. diameter are generally in steps of $\frac{1}{16}$ in. If the diameter of an old rope is found to be $2\frac{3}{32}$ in., this dimension is $\frac{1}{32}$ in. less

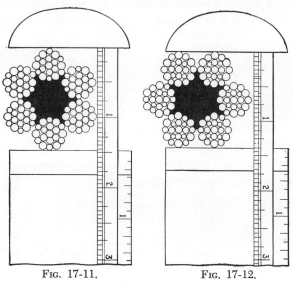

Fig. 17-11.   Fig. 17-12.

Fig. 17-11.   Correct diameter of wire rope.
Fig. 17-12.   Incorrect diameter of wire rope.

than $\frac{3}{4}$ in. Therefore it will be safe to assume that the rope was $\frac{3}{4}$ in. in diameter when new. For elevator service the sizes most commonly used for hoisting purposes are $\frac{1}{2}$ and $\frac{5}{8}$ in. in diameter or larger. Operating and governor ropes are usually $\frac{1}{2}$ in. or less in diameter. It is important that the correct size be selected, since an oversized or undersized rope will have a short life.

Ropes Used on Drum-type Elevators. Present tendencies seem to be toward the use of either the 8×25 filler wire or the 8×19 Seale types. Iron ropes of the 6×19 regular or Warrington constructions at one time were used almost exclusively for hoisting, counterweight, and governor service on drum-type machines. They are still in use, but for drum-counterweight ropes on basement-type installations 8×19 Seale is also used with satisfactory results. On these machines the counterweight ropes have a reverse bend in them as they lead off the drum and up the hoistway (see Fig. 2-1). In some cases this bend is short, but even under the most favorable conditions a reverse bend fatigues the wires in the rope faster than when all bends are in the same direction. Experience with 6 × 37, 8 × 19, or 6 × 25 filler mild-steel ropes for these applications has shown good results. Mild-steel ropes of the 6 × 19 construction are sometimes used for car-hoisting service.

Ropes Used on Traction Elevators. In general, drum-type elevators are used on comparatively short rises, and run at slow speeds in service where they are not in continuous operation. As a result car mileage per year is comparatively small and the ropes may operate for 10 years or more before replacement is necessary. For traction machines conditions are in general quite different from those for the drum type. Cars frequently operate at high speeds and are in continuous operation for 9 or 10 hours per day; consequently car mileage per year is high and ropes wear accordingly. Traveling 10,000 miles or more per year is not uncommon for high-rise express cars. To increased wear and fatigue due to high yearly car mileage must be added effects of slippage and wedging of the ropes in the traction-sheave grooves. These conditions impose severe service on the ropes of traction machines.

For this service mild-steel ropes of various constructions are used, iron being unsuited. Mild-steel ropes are sold under various trade names, such as traction steel, special traction steel, toughened steel, etc. All, however, are made of a steel specially treated to make it suited to the service conditions existing on traction elevators. Warrington-type ropes have been used extensively on traction machines, but there is a tendency toward more flexible constructions. One manufacturer recommends, for double-wrap traction machines with sheaves over 29 in. in diameter, a 6 × 19 with six small spacer wires in each strand, making a 6 × 25 construction. For double-wrap traction machines with sheaves under 30 in. in diameter, all V-grooved-type traction machines, and machines with 2-to-1 roping, an 8 × 19 Seale construction is recommended. All these ropes are made of special traction steel. These recommendations are similar to those of another manufacturer making a toughened-steel rope.

Practice appears to lean toward the use of more flexible ropes, such as the 8 × 19 and 6 × 19 Seale constructions. These give a flexible construction with large wires on the outside of the strands to stand wear, and some manufacturers recommend this construction for practically all traction elevators.

Governor Ropes. At one time practically all governor ropes were 6 × 19 or 6 × 42 iron. Mild-steel ropes of the 6 × 19 and 8 × 19 constructions are now used extensively. The tail rope, connecting the drum on wedge-clamp-type safeties to the governor rope, is frequently made of bronze. This piece of rope is seldom in use and, if not looked after, may become badly rusted and weakened. Using bronze rope for this service eliminates the rust hazard so far as the rope is concerned. Of course, if the safeties are allowed to become rusted so that they cannot be applied, a bronze tail rope is of little use.

Compensating Ropes. For compensating ropes both 6 × 19 and 8 × 19 steel or iron constructions are used.

Regular-lay Right-lay Construction. Regular-lay right-lay is the construction usually recommended for elevator service. Although Lang-lay rope offers a greater wearing surface to the sheave grooves, it is more difficult to handle than regular lay, and has not come into general use for elevator service.

The strength of the ropes does not usually come into the problem of selection after an elevator has been installed. As the equipment is usually designed, the safety factor of the ropes is above that recommended by The American Standard Safety Code for Elevators, Dumbwaiters and Escalators. These factors of safety are shown by the curves in Fig. 17-13. Tags on the elevator indicate the rope strength that should be used.

Causes of Reduced Wire-rope Life. Elevator rope life depends upon a multitude of factors and may extend over periods ranging from a few months to 15 or 20 years. Although it is quite general practice to express life of wire ropes in terms of the time they have been in service, this method is not so accurate a measure as car mileage. In many of the better-operated buildings recorders are installed on the elevators and a record kept of car mileage, and rope life is based on these records. Car mileage is not, however, always a measure of rope mileage, since on 2-to-1 roped traction machines the rope speed is twice car speed. With such installations it is necessary to multiply the car mileage by two to obtain the hoisting-rope mileage, which is the measure of the service rendered by these ropes.

There are many ways of roping up elevator machines, as shown in Chap. 2, and they all influence rope life. Other conditions being equal, the basement-type traction (Fig. 2-9) and the 2-to-1 traction roping (Fig. 2-10) will have the shortest life, because of the number of bends and reverse bends in the ropes.

On basement-type drum machines the counterweight ropes come

off the drum and make a reverse bend around a sheave to go up the hoistway (Fig. 2-1). In low-ceiling basements it is necessary to place the sheave near the drum, so that the ropes bend almost 90°. Such a reverse bend tends to shorten rope life and should be avoided where possible.

Wire ropes wear out from two causes: 1, a reduction in area caused by wearing of the outside wires; and 2, breakage of the wires. In elevator service the latter is the reason for condemning most elevator

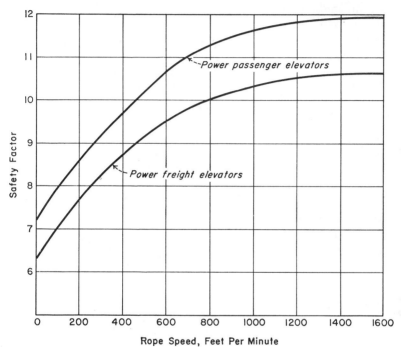

FIG. 17-13. Factors of safety for hoisting ropes, based on the 1955 American Standard Safety Code for Elevators.

hoist ropes. From this the conclusion might be drawn that there has been a tendency to use elevator ropes that are too stiff for the service. Now the trend is toward the more flexible 6 × 19 Seale, and 8 × 19 instead of the regular 6 × 19 constructions almost exclusively used at one time.

Relation of Drum or Sheave Diameter to Rope Diameter. It is recommended that sheave or drum diameter be as large as possible, consistent with other conditions. It is preferred not to have traction sheaves less than about 40-rope diameters. Larger diameters than this are favorable to rope life.

Conditions of the sheave grooves have a marked effect on hoisting-rope life, particularly on traction machines. If the grooves are worn to fit old ropes, a new set will wedge into the grooves. This wedging action tends to distort the ropes and cause fatigue and breakage of the wires. Since the grooves are generally not all worn to the same tread diameter, some ropes tend to travel faster than others. To accommodate this difference in rate of travel some of the ropes must slip in the sheave grooves. If they wedge into the grooves, the difference in loading of the ropes before slippage occurs will be increased and may cause excessive stresses to develop in some of them. As the surface wires must resist the forces that cause slippage, they may be stressed until actual breakage occurs.

When the hoisting ropes are to be changed on a machine, check the sheave grooves. If they are found to be undersized or of unequal tread diameter, put them in proper condition. If more attention were given to this feature of elevator operation, the life of hoisting ropes would be greatly increased in many cases.

Equalizers on Traction-Elevator Hoist Ropes. Equal tension should be maintained in the hoisting ropes. This is difficult, since one rope may have a tendency to stretch more than another or the grooves in the sheave may not have exactly the same tread diameter. At one time it was quite general practice to use automatic equalizers on the hoisting ropes of traction machines, but this practice is losing favor. Experience has shown, however, that unequal loading of ropes due to excessive stretch of some of them has been the cause of short rope life in more cases than is generally supposed.

It is necessary for the ropes to lead off the sheave without rubbing one side of the grooves. This applies particularly to the sheaves on drum-type machines. One of the bearings worn on an overhead sheave would throw it out of line. If the shaft on which the vibrating sheave travels is not kept well lubricated, the sheave will lag on the shaft and throw the ropes out of line with the grooves as they come off the drum, causing unnecessary wear on the ropes.

Ropes on a high-rise car will have a longer life in car miles than those on a low-rise car. This is to be expected, since the high-rise car makes fewer trips per car mile, which means less bending of ropes around the sheaves and less wear on them.

When Ropes Should Be Renewed. Another factor in rope life is the judgment of the man who condemns them. There is no method by which the condition of a set of ropes may be exactly appraised. A general rule was to condemn the ropes when six wires were broken in one lay of one strand in the worst section of the rope. Recently the National Elevator Manufacturing Industry, Inc., 101 Park Avenue, New York City, made the following recommendations for single- and double-wrap traction machines that have three to eight hoist ropes:

1. If the broken wires are equally distributed among the strands, remove the ropes when the number of broken wires per rope lay in the worst rope section exceeds 24 to 30 in a 6×19 and 32 to 40 in an 8×19 rope. A wire-rope lay is the distance along the rope within which the spiral strands complete one turn around the rope's core. A lay may be considered as a length equal to about $6\frac{1}{2}$ rope diameters; that is, $3\frac{1}{4}$ in. for a $\frac{1}{2}$-in. rope and $4\frac{1}{16}$ in. for a $\frac{5}{8}$-in. rope.

2. If broken wires predominate in one or two strands, remove the ropes when the number of broken wires per rope lay in the worst rope section exceeds 8 to 12 in a 6×19 and 10 to 16 in an 8×19 rope.

3. If four or five adjacent wires are broken across the crown of the strands, remove the ropes when the number of broken wires per rope lay in the worst rope section exceeds 12 to 20 in a 6×19 and 16 to 24 in an 8×19 rope.

4. Under unfavorable conditions, such as corrosion, excessive wear of individual wires in the strands, unequal tension, poor sheave grooves, etc., remove the ropes when the number of broken wires exceeds 50 per cent of the numbers given.

For drum machines with two hoist ropes, if broken wires are equally distributed among the strands remove the ropes when the number of broken wires per rope lay in the worst rope section exceeds 12 to 18. If broken wires predominate in one or two strands, remove the ropes when the number of broken wires in the worst rope section exceeds 6 to 12.

Note in all cases that the higher values may be used only when the ropes are inspected at least once a month by a qualified inspector.

It is also recommended that ropes be replaced when the diameter of a $\frac{1}{2}$-in. or a $\frac{9}{16}$-in. rope drops $\frac{1}{32}$ in. below its rated value; a $\frac{5}{8}$-in., a $\frac{11}{16}$-in., or a $\frac{3}{4}$-in. rope drops $\frac{3}{64}$ in. below its rated value, and when the diameter of a 1-in. rope drops to $\frac{15}{16}$ in.

Attention is also called to the fact that broken wires are difficult to detect in preformed ropes since these wires do not spring out when broken. When inspecting preformed ropes clean them with a solvent and use adequate lighting.

Putting Sockets on Wire Rope. The method generally used by elevator mechanics for socketing elevator ropes is shown in Fig. 17-14. First the rope must be seized with annealed iron wire to hold the rope lays in place. Make each seizing one rope diameter wide.

Seize nonpreformed rope at three places, each side of where it is to be cut. Put the first seizing close to where the rope is to be cut and the second back from the first the length of rope to be turned in. Place the third seizing a distance from the second equal to the tapered length of the socket.

After the rope has been seized and cut, push the end to be socketed through the small end of the socket and up on the rope, as at *B* in Fig. 17-14. Remove the two end seizings, fan out the strands as far as the remaining seizing, and cut out the fiber core as close as possible to the seizing. Clean the exposed strands of grease and oil with a nonflammable solvent. Each strand should then be bent back and

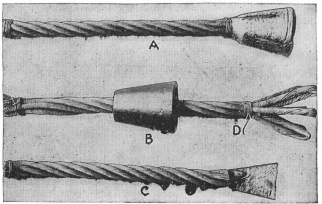

Fig. 17-14.   Turned-in method of socketing elevator hoist ropes.

inward on itself the same distance, bringing the ends close to each other in the center, as at *D* in Fig. 17-14. The turned-in ends should be about 2½ rope diameters in length. When pulled as far as possible into the socket the bends of the turned-in strands should extend slightly beyond the mouth of the socket and be visible when the socket is babbitted. Take care to make the bends as nearly equal as possible and avoid driving them into the socket, as this may distort the rope and cause high strands, on which excessive wear may occur.

Before pouring babbitt into the socket place it in a vertical position with the large end up and be sure that the rope's axis is parallel with that of the socket. Close the small end of the socket by winding waste or tape around the rope to hold in the babbitt until it cools. Heat the socket with a blowtorch sufficiently to prevent chilling the babbitt and to ensure that the babbitt will fill the socket and the space between the rope strands.

Heat the babbitt until it will just char a piece of soft wood and slowly pour into the socket until it is full. After the babbitt has cooled remove the tape or waste from the small end of the socket to see that babbitt is visible here. Move the socket back on the rope to see that all strands are well imbedded in babbitt as at *A* in Fig. 17-14. At *C* the terminal is cut through parallel to the rope's axis and shows the strands and babbitt forming practically a solid mass,

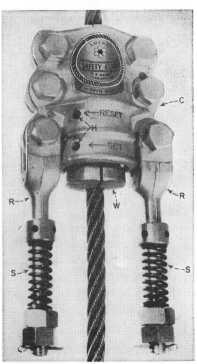

Fig. 17-15.   Safety clamp connects to hoist ropes at car crosshead on drum-type elevators.

as they should when the job has been done properly as recommended by The American Safety Code for Elevators. Other methods of socketing wire ropes are not recommended for elevator service.

Periodic Resocketing.   Because the hoist ropes on drum-type elevators have fatigued and failed unexpectedly at the car-crosshead sockets, some local authorities require periodic resocketing of these ropes. The American Standard Safety Code for Elevators, Dumbwaiters and Escalators recommends that when drum machines are installed overhead the hoisting ropes be resocketed at the car every 12 months.

For basement-type machines this period is increased to 24 months. When resocketing the ropes cut a sufficient length from the ends of the ropes to ensure that all damaged or fatigued portions are removed and that there is no appreciable change of rope lays immediately above the cut-off end. A metal tag should be securely attached to each rope after resocketing to show who did the job and the date.

Safety Clamps. In lieu of resocketing, auxiliary rope-fastening devices may be installed, such as a Lucas automatic safety clamp (Fig. 17-15). This device consists of a case C and a wedge W drop forged in two identical halves so that existing rope attachments need not be disturbed for installation. Case C is mounted on the car crosshead above the hoist-rope sockets and bolt connections, by yoke bolts and extension rods R. Compression springs S support case C over the free-floating wedge W, which bolts directly to the rope inside case C but is not in contact with it, allowing the rope its original free movement.

Until the rope fails at its car-crosshead socket there is no contact between the case and the wedge. When a rope fails at its socket, wedge W moves up against case C as the weight of the car is taken by the clamp. The maximum drop of 1.25 in. is so slight as not to disturb passengers in the car.

Holes H permit the inspector to see the condition of the rope at its socket easily. When the clamp is installed a red line on the wedge is matched with the lower hole in the case. When the line leaves its set position it gives warning of pending failure of the rope, which can then be resocketed before failure occurs.

CHAPTER 18

Electrohydraulic Elevators[1]

About 40 years ago, hydraulic elevators reached their acceptance peak for low- and high-rise applications. Then they were gradually superseded by electric types. One hydraulic design required a plunger casing that went into the ground as far as the car went up; in some buildings 30 stories or more. Many such machines are still operating, although frequently with modern electric control in place of their original mechanical scheme.

Modern Hydraulic Types. The operating principle of the old plunger elevators is again being applied in many of today's lifts and freight and passenger elevators. Figure 18-1 shows all the elements in a simple modern plunger hydraulic elevator. Oil, the power fluid, is stored in a tank from which it is pumped through a check valve into the cylinder to lift the plunger and car with its load. The cylinder-and-plunger assembly is frequently called a jack. The power unit is a motor-driven positive-displacement pump started and stopped by the up-direction push button.

Three valves are required, the first being a check valve to oil-lock the car in position when the pump is stopped by releasing the up push button. A relief valve connects from the pump's discharge into the return line to avoid producing excessive pressure in the system if the car is overloaded. In the smaller installations this valve may be built into the pump, between its discharge and suction. A solenoid-operated valve functions to lower the car. In this way, oil under pressure lifts the car and its load and gravity lowers them.

Closing the up button starts the motor and pump to discharge oil from the reservoir to the casing or cylinder. When the car reaches the floor the up-direction button is released, the pump stops, and

---

[1] Electrohydraulic elevator companies supplying data for this chapter besides those mentioned in the text are: S. Heller Elevator Company, Murphy Elevator Company, E. Rosenberg Elevator Company, White Evans Elevator Company, and Esco Elevators, Inc.

the check valve closes to oil-lock the car at the floor. Pressing the down button opens the solenoid valve to lower the car by letting oil flow from the plunger cylinder into the reservoir. When the car comes to the landing, releasing the button closes the solenoid valve to stop the car at the floor.

As Fig. 18-1 shows, this type of elevator combines the old hydraulic-plunger lift with an electric-motor driven pump and an electrohydraulic control. Even though it is called hydraulic and by many trade names, such as oildraulic, oil-lift, oilelectric, ramlift, and others, probably the most appropriate is electric-hydraulic or electrohydraulic. These elevators will be called electrohydraulic in this chapter, as designated by The American Standard Safety Code for Elevators.

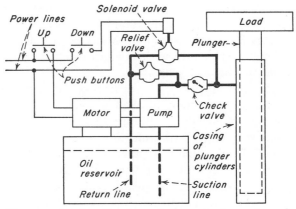

Fig. 18-1. Diagram showing the elements in a simple modern plunger electrohydraulic elevator.

All electrohydraulic elevators and their controls are enlargements of that shown in Fig. 18-1, with added safety devices, bypass and leveling valves, and other refinements to obtain a safe, smooth, and efficient operating machine. Using oil as a pressure fluid instead of water eliminates many troubles with water such as rust, corrosion, and dangers of freezing.

Cylinder and Plunger Designs. The single-bearing cylinder and plunger, shown at *A* in Fig. 18-2, operates like a hydraulic jack. Liquid under pressure forces the plunger up, lifting the load. A cylinder or outer casing of heavy-steel tubing is generally sunk in the ground as far as the car rises. A plunger, also of thick-walled steel tubing and polished to a mirror finish, connects to the car. It is sealed in the top of the cylinder with compression packing *K*.

Babbitt-lined steel bearing L reduces friction and prevents the plunger from scoring. Oil under pressure is admitted near the top of the cylinder through connection F. Bleeder G removes air from the system, preventing jerky operation of the lift. Stop plate Q welded to the plunger's bottom prevents the plunger from being forced out of the casing. At its top end, a welded bulkhead fastens the plunger to the car sling or to the bottom struts of the lift.

Off-balance Loads. Cylinders having a single bearing at the top are suited for elevator and sidewalk lifts where the car is guided top and bottom. Guides prevent off-balance loads from forcing the plunger off vertical and placing a side thrust on the cylinder's bearing. Where the platform is not guided, as is true on many lifts, several cylinder designs are used to give the plunger greater support against off-balance loads.

One such cylinder (Fig. 18-2B) has two widely spaced steel rings R electrically welded into a thick-walled steel barrel S. Babbitt faces in these rings take side thrust, giving smooth operation and preventing scoring of the plunger. Both bearings always operate immersed in oil. The top of the cylinder is sealed with square and round hollow-core compression packing K. Under ordinary conditions the hollow core automatically adjusts pressure on packing to compensate for wear.

Movable Bearing. Other fluid cylinder and plunger designs for off-center loads are shown in Figs. 18-2C to 18-2E. The design shown in Fig. 18-2C may be used for general industrial applications since its heavy-service plunger and cylinder walls will stand stress of heavy off-center loads. Movable bearing M, fitted to the lower end of the plunger, gives added support against such loads. The plunger is supported by a removable bearing R at the cylinder top that contains the packing. Five rings of chevron packing P, held under gland G, seal the plunger in the cylinder. Oil under pressure is admitted at F, flows down between the plunger and cylinder wall, through holes H, then under the plunger to lift the load.

Cage Bearing. In Fig. 18-2D the design has a cage bearing R supported by cylinder S about 3 ft down from the main cylinder head. Oil under pressure enters at F, passes down through holes in the bearing, then below the plunger to lift the load. This design is adapted for long-stroke service where the platform rise is higher than usual, as in loading, production, and bridge-lift service.

Double-acting Cylinder. Where the platform is not heavy enough to ensure gravity lowering, a double-acting cylinder (Fig. 18-2E) may be used. It is also applied where the cylinder must be mounted

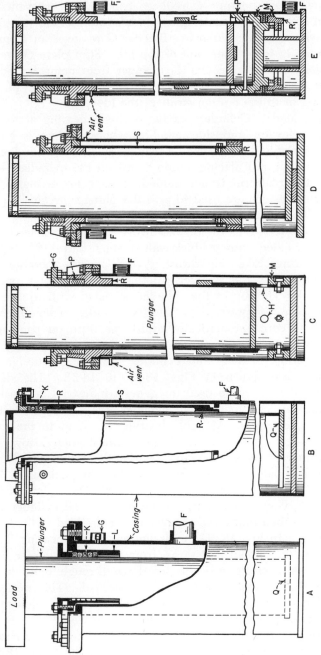

Fig. 18-2. Types of hydraulic elevator jacks. *A,* single-bearing plunger for guided loads. *B,* two-bearing plunger for off-balance unguided loads. *C,* a movable-bearing plunger for heavy service. *D,* cage bearing for long-stroke service. *E,* double-acting plunger.

horizontally for power operation on out and return strokes. The plunger is sealed in the cylinder's top as shown in Figs. 18-2C and 18-2D. Piston P fastens to the plunger's bottom and is sealed in the cylinder by two mechanical packings M, preventing leakage in either direction.

A short section of pipe fastened to the bottom of the cylinder prevents the load from landing on bull ring R_1, which holds the packing in place. Ring R is welded to the plunger and stops the piston before the bottom bearing strikes the cylinder head. To lift the load, pressure oil is admitted below the piston at F. When the platform is lowered, pressure oil flows in at F_1 above the piston and out at F below. There are other jack designs, but those described give a

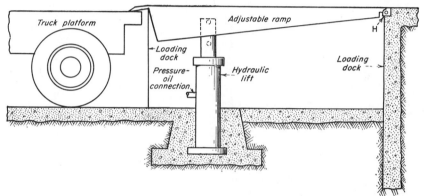

FIG. 18-3. Adjustable ramp in a loading dock has its near end hinged to the dock at H, while its front end is connected to a hydraulic lift through a swivel bolster.

good general idea of the types used. Jack-plunger sizes range from 2.5 in. in diameter on small low-capacity lifts to 18 in. on large heavy lifts, operating at 150 to 400 psi. This range of plunger sizes and pressures takes care of most load requirements.

Applications. One of the most common hydraulic-lift applications is that in automobile service stations. The same type of lift is used in many industrial plants. A typical industrial application would be an adjustable ramp on a loading dock (Fig. 18-3). This ramp hinges to the dock at H, while its front end connects to a hydraulic lift through a swivel linkage bolster, bolted to the ramp's underside.

The outer end of the ramp is flush with the side of the dock, but is raised above truck-bed level before the truck backs in. After it is in position, the ramp is lowered until the bridge plate rests on the truck's bed (Fig. 18-3). Pressure is then relieved under the plunger, after which the ramp follows up or down with the truck bed as load is put on or removed. This permits a fast, smooth flow of material on or off the truck. The hydraulic cylinder sets in the floor under the ramp.

Platform Lifts. Of the many types of hydraulic lifts available, one has the platform alone mounted on a hydraulic plunger (jack), as shown in Fig. 18-4. Such an installation could be used to: (1) handle freight from a plant floor into trucks, freight cars, or different building levels; (2) position material for machine feeding; or (3) lift heavy loads in a host of possible applications.

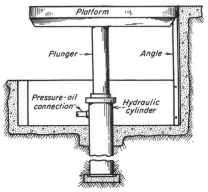

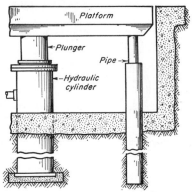

Fig. 18-4. This hydraulic lift consists of a platform only, mounted on the plunger of the jack.

Fig. 18-5. Pipe may be used to prevent platform rotation when the lift is in the center of the plant's floor.

The platform of a single-plunger lift can turn freely with the plunger, so it must be kept from rotating during up and down travel. The inner corners of the platform (Fig. 18-4) run in a pair of heavy angles extending from the lower to the upper landing. When there is no adjacent wall to fasten angles, as when the lift is installed in the center of a floor, a pipe may be used to prevent rotation (Fig. 18-5). Even though the pipe keeps the platform from turning, it does not relieve the plunger and cylinder of side thrust or loading shocks. For maximum effectiveness, locate the pipe as far as possible from the jack.

Dual Plungers. Several jacks may be used to obtain maximum platform stability (Fig. 18-6). Seldom is it necessary to use more than two jacks, but four have been used on a circular stage lift. To keep the platform level with only two jacks under all loading conditions calls for an equalizing system. Typical are channel-iron struts, shown at *A* in Fig. 18-6, that run from the platform near the top of each plunger down into the jack well.

A bearing support for two double-grooved sheaves, S and S_1, connects to each cylinder. One pair of wire ropes anchored to the top of the right strut drops down and goes under sheave S, up and over sheave S_1, then down again to the bottom of the left strut. The other two ropes anchor to the top of the left strut, come down and under S_1, then up and over S sheave to the bottom end of the right strut.

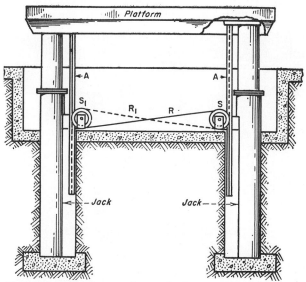

Fig. 18-6.   Two or more jacks may be used when a single hydraulic cylinder and plunger fail to give sufficient stability to the platform.

When load on the right side of the platform is greater than on the left, the platform tends to get out of plumb and the left end tries to rise. But the lower end of the right strut connects to the top end of the left through ropes R. These ropes form a mechanical tie between the bottom of the right strut and the top of the left one. As a result, the right end of the platform cannot go down without also pulling down its left end.

Now let's reverse load so that the left end of the platform has the greatest load. Note that this end cannot go down without pulling the right end down with it, since rope R_1 serves as a mechanical tie between the platform ends. In brief, the equalizer ropes hold the platform level under all loading conditions. At no time is the equalizing system loaded more than one-half the difference in load on both ends of the platform.

The hydraulic system of dual jacks consists of a single power unit with a pipe leading to a manifold. From this manifold a pipe runs to the top of each jack cylinder. Otherwise, both power unit and control system are the same as for a single jack used for the same purpose.

Other Equalizers. Another equalizer system uses gears rather than ropes. Platform ends are connected through gears, one at each end of a shaft supported in two bearings and extending from one jack to the other. A strut with rack teeth connects to the platform at each

Fig. 18-7. Adjustable loading ramp with hydraulic lift on which are assembled the motor, pump, valves, and piping, to form a compact unit.

jack, as in Fig. 18-6. Shaft gears mesh into the rack-gear on each strut. The equalizer ties the lift ends together so that one cannot move unless the other moves in the same direction and at the same rate. A third type of equalizer meters fluid to the jacks so that each moves at same rate, keeping the platform level.

A feature of the lifts we have studied is their extreme simplicity and their easy adaptability to a variety of low-rise jobs. All working parts are below floor level, so that when the lift is lowered, its platform forms part of the floor without any obstructions above. On some smaller lifts, the pumps, valves, and piping are assembled on the jack (Fig. 18-7). With this arrangement the only equipment above floor level is the push button to start and stop the pump motor.

Freight and Passenger Elevators. Modern electrohydraulic freight or passenger elevators are a highly refined development of the hydraulic lifts we have been considering. Figure 18-8 shows a passenger type, but freight elevators have the same general design except that they can handle heavier loads and take greater shocks and side thrust. These elevators differ from electric designs in several ways. The first is in their extreme simplicity. The car rests on the hydraulic plunger, by which it is lifted. There are no wire ropes, therefore, no overhead sheaves or penthouse.

Since there is no overhead equipment, hoistway columns and footings may be of lighter design than for a conventional electric elevator. Car safeties or speed governors are not needed because the car and its load cannot fall faster than normal speed. Hydraulic elevators travel at low speeds, so the pit under the car can be quite shallow and the bumpers need be only heavy springs.

Electric equipment in the hoistway and on the car is not shown in the diagram, but it too is quite simple, consisting chiefly of wiring to the car push buttons and lights.

Elevator Ratings. The capacity of passenger elevators ranges from 1,000 to 4,000 lb at speeds of from 40 to 125 fpm. Since electrohydraulic elevators lower by gravity, down-speed may be 1.5 to 2 times up-speed with the help of simple valve adjustments. Thus, average speed for a round trip can be considerably higher than up-speed. For example, if up-speed is 60 fpm and down-

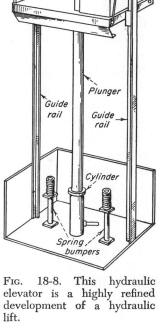

Fig. 18-8. This hydraulic elevator is a highly refined development of a hydraulic lift.

speed 90 fpm, the average is $(60 + 90)/2 = 75$ fpm. This feature is inherent in the equipment.

Standard freight-elevator capacities run from 2,000 to 20,000 lb at 20 to 85 fpm, but units have been designed and installed for 100,000 lb or more. Rises to 50 ft, adequate for two- to five-story buildings, are the most practical and economical for freight and passenger service. Within this rise range, casing and plunger can be made and installed in one piece—far simpler than the assembling of these parts in sections. For higher lifts, car speed may be greater, but with increased power needs and more complicated control.

For freight-elevator maximum economy and top service, an up-travel time lapse of 30 sec between floors is recommended for a three-floor rise where the unit is not part of a production line. This means travel from first to third floor takes 1 min. Where faster speeds are needed, equipment is available with speeds of up to 125 fpm. However, higher speeds require greater pump capacity, more motor power, and more complicated control, materially adding to costs of elevator installation and operation. For short-run passenger elevators a time lapse of 15 sec between floors is average service, but this, too, can be reduced to about 5 sec or 120 fpm.

Cars Available. Passenger and freight cars for electrohydraulic elevators are comparable with electric designs in every way. They will meet requirements of the most rigid safety codes.

Passenger elevators can be equipped with any entrance design specified. Frames, sills, doors, trim, and door-hanger covers are available in a wide range of patterns, metals, finishes, and colors. Elevator-door designs can be simple or ornate, modern or traditional, to match building interiors.

Air Hydraulic. Power units and their control form the heart of all electrohydraulic elevators and lifts. Probably the simplest are those used with manually controlled air-hydraulic automobile lifts in garages. Figure 18-9 diagrams one design of air-oil mechanism used chiefly for automobile lifts. A compressed-air line comes into the top of the cylinder casing, drops to the bottom, then goes up into the plunger. The casing is filled with oil and the plunger bottom is closed except for clearance holes for an air pipe and control valve.

If the casing is filled with oil, the air valve needed to operate the lift is less than 50 per cent of that required without oil. Oil also shortens the air column, making the unit less likely to change level with load. Further, oil serves as a safety device, since the lift cannot rise or lower faster than the time it takes the oil to flow through the control valve in the plunger's bottom. The lift could be raised fast at light loads and dropped fast at heavy loads were it not for the oil-control valve in the plunger's bottom. The control valve holds the speed to a safe value in either direction of travel.

In the unit shown in Fig. 18-9, a lever-operated valve controls the flow of compressed air to and from the jack's casing. Air is ad-

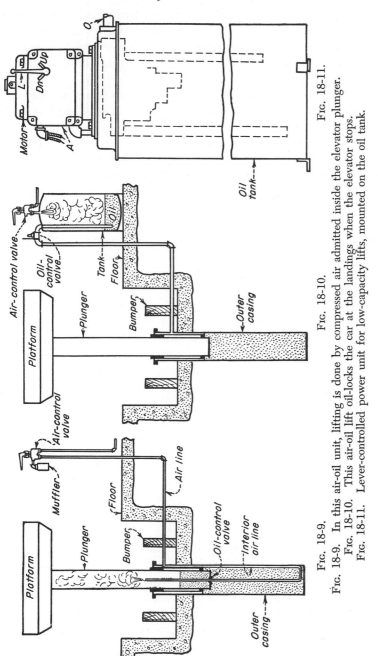

Fig. 18-9.

Fig. 18-10.

Fig. 18-11.

Fig. 18-9. In this air-oil unit, lifting is done by compressed air admitted inside the elevator plunger.
Fig. 18-10. This air-oil lift oil-locks the car at the landings when the elevator stops.
Fig. 18-11. Lever-controlled power unit for low-capacity lifts, mounted on the oil tank.

mitted to the plunger by lifting the control-valve handle. Pressing down the lever opens the discharge valve in the control unit, releasing air through the muffler, and thus allowing the load to lower. With the lever in neutral position, both the admitting and discharge valves close, stopping the lift and its load.

Since air is elastic the lift hunts up and down slightly as load changes. It is mainly for this reason that the unit has limited application outside of automobile lifts. The system is used to a limited extent in industry for machine-feed applications where a slight upward float helps keep load level with the feed table.

Oil-lock System. Compressed air is the power medium, with oil making direct contact with the plunger in the system (Fig. 18-10). With the lift at the lower landing, most of the oil is in the pressure tank. Atop the tank is a compressed-air control valve. An oil control valve is in the line to the plunger's cylinder. With the compressed-air admission and oil valves are opened, air pressure forces oil from the tank into the casing around the plunger, raising the load as shown in Fig. 18-10.

Releasing the dead-man oil control valve locks the system, leaving the load supported on the oil column in a fixed position. To lower the lift, air is exhausted from the tank through a muffler, not shown in Fig. 18-10, and the oil valve is opened.

Operation of hydraulic lifts with compressed air is usually practical only when an air supply of 90 psi or higher is available. Even then it is recommended for limited use, generally when less than five trips are made per hour, or speed is not a major factor.

It is seldom economical to use compressed air for large lifts. Air cost is about 10 times that of power taken by an electric motor driving an oil pump. Once again, where operating speed is a factor, don't use compressed air. Even though the plunger will move at a speed comparable to that of an oil-powered unit, considerable time is lost in charging and discharging air when moving the lift.

Outside of the air-oil lift applications mentioned, most hydraulic lifts and elevators are operated by an oil-power unit. These units consist of a motor-driven pump, an oil reservoir, and a system of valves for controlling oil flow to and from the plunger casing. Pumps used in these units are generally the positive-displacement type, of either the rotary or rotary-piston designs.

Small Power Units. A simple lever-controlled power unit for low-capacity lifts is shown in Fig. 18-11. It consists of a vertical motor driving a gear-type rotary pump mounted in the top of an oil tank. Valves, controlling oil flow to and from the hydraulic cylinder and lift plunger, are also in the top of the tank. Control assembly *A*, actuated by lever *L*, is mounted alongside the motor. The pump is rated 28 gpm at 185 psi, with a tank capacity of 32 or 75 gal depending on lift rating. Moving the lever to the right starts the pump and opens the oil-intake and discharge valves. Oil is discharged through pipe *O* to the cylinder to raise the plunger and load.

When lever *L* is released it automatically comes to the center or neutral position, stopping the pump, and the check valve closes to oil-lock the lift in position. Moving the lever to the left opens a

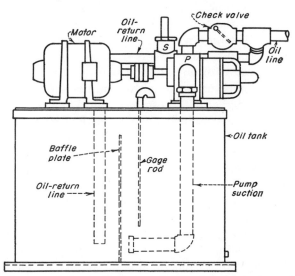

Fig. 18-12. Power unit for medium-capacity lifts has constant-speed motor driving a rotary pump with control valves mounted on the pump.

valve, to let oil flow from the plunger cylinder back into the tank through pipe *O*, as the lift lowers.

Hydraulic equipment, for a larger lift than those discussed, is shown in Fig. 18-12. It consists of a horizontal-shaft constant-speed motor coupled to a gear-type rotary pump *P* having a built-in pressure-relief valve. The pump discharges through a check-valve into an oil line leading to the lift's hydraulic cylinder. This check valve oil-locks the lift in the position it is in when the pump stops.

This arrangement of an electric-motor-driven pump and valves is basic to all hydraulic-lift and elevator power units. On all power units the pump discharges through a check-valve. In that way, when the pump stops, the lift or elevator car is oil-locked in position. A solenoid valve *S* bypasses the check valve to lower the car or lift. The solenoid-valve opening is sized to limit down movement of the lift or elevator car to a safe value. A pressure-relief valve is either built into the pump or installed in the connecting piping.

Limit Switches. On all electric elevators, limit switches and other safety devices must be provided to prevent the car or counterweight from being pulled into the overhead structure. If such an accident were to happen on a drum-type machine, a serious wreck would result. Limit switches guard against such accidents. However, hydraulic lifts and elevators have no such hazards.

As an example, the hydraulic lift does not require terminal-floor limits. Speed is so low that even if the operator ran the lift to the limit of up-travel, the plunger bottom would contact the cylinder head and stop. Then, even if the pump continued to run, the relief valve would open on a small excess pressure, bypassing the oil back to the storage tank, with no resulting harm. If the operator is daydreaming and permits the car to go beyond normal limits, pump noise and relief-valve popping will rouse him quickly.

If the solenoid valve is held open when lowering, the loaded lift can only go down until it rests on the bumpers or the plunger hits cylinder bottom. In either case, the speed is so low that no harm is done. For these reasons, control of hydraulic lifts may be stripped to the bare essentials.

Control Refinements. For passenger- and freight-elevator service the control requires a few more refinements than are needed for simple lifts. Such a power unit, as built by the Globe Hoist Company, is shown in Fig. 18-13. A triple-screw pump *P* driven by a squirrel-cage motor *M* is used for quiet and smooth operation. This pump produces a smooth oil flow without objectionable pulsations. Both pump and motor are mounted on a heavy structural-steel base *B*, separate from the oil-storage tank. A rubber coupling and rubber sound-insulating mounts reduce noise transmission from the motor and pump.

On this unit suction *S* is on the left of the pump and the discharge is on the right. A 5-ft length of flexible tubing *F* connects the pump's discharge to the solid-piping manifold, which includes the control and other valves. Pipe *C* runs to the elevator's hydraulic cylinder. Check valve *V* oil-locks the ·elevator car in position. In case of overpressure, relief valve *R* bypasses oil back into the oil-return line and the tank at *T*. The oil-return line connects into the hydraulic cylinder line beyond the check valve. From there the oil travels up through a strainer *E*, to down control valve *D*, and into the tank at *T*. Note that the return-line connection into the cylinder line is hidden behind relief valve *R*.

Down valve *D* is bypassed by leveling valve *L*. This valve is controlled by a switch near the landing actuated by a cam on the car. Accelerating valve *A* serves to accelerate the car smoothly upward by relieving the shock of the pump suddenly starting.

Starting and Stopping. The power unit is controlled by push buttons at the landings or in the car. Pressing an up button starts the motor and accelerates the pump. When the pump starts, part of its discharge is bypassed back through valve *A* to the oil reservoir. As

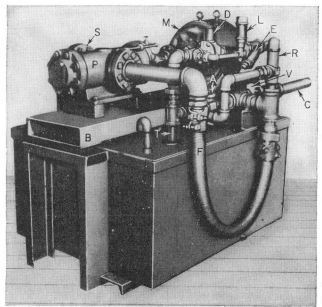

Fig. 18-13. Passenger-elevator power unit has a quiet-operating triple-screw rotary pump supplying a smooth flow of oil to the lifting jack.

this valve closes slowly, oil is directed into the hydraulic cylinder and the car accelerates smoothly to rated speed. When the pump is stopped, the car comes to rest slightly above the floor to which it was signaled. A car cam then closes the leveling switch. This switch opens the leveling valve, permitting oil to flow slowly from the jack cylinder. Flow from the cylinder brings the car back to the floor, where the cam releases the leveling switch, closing the leveling valve.

Down Operation. When a down push button is pressed, control valve *D* opens and lets oil flow from the jack back to the reservoir. The car and its load lower by gravity until the down valve closes. Control for this valve is arranged so that it closes when the car is slightly above the floor at which it is to stop. Then the leveling valve opens, bringing the car level to the floor when the leveling valve closes.

Valve Assembly Unit. Figure 18-14 shows the valve assembly unit for Rotary Lift electrohydraulic elevator controls. The unit carries out the instructions from the electric control panel, actuated by the

Fig. 18-14. Valve assembly unit responds to signals from an electric control panel actuated from push-button stations.

car and landing push buttons. It combines eight control valves. External piping consists of P leading to the oil-storage reservoir, and P_1 going to the jack cylinder. All valves are externally adjustable without the unit being removed from the system.

The valve unit is mounted directly on the suction and discharge of a rotary-piston pump (Fig. 18-15). The complete assembly is enclosed in a steel housing with an oil-storage tank on top. Cabinet E houses the electric control panel. It may be mounted on either end of the steel housing or on an adjacent wall. The pump is driven by a squirrel-cage motor *M*, V-belted to a large sheave *S* on the pump drive shaft.

Units of the above type are built in ratings of 5 to 80 hp for freight elevators handling loads to 100,000 lb, with car speeds of up to 100 fpm. For lighter-duty lower-speed elevators, a motor-driven helical-gear pump may be used. It is incorporated in power units of

Fɪɢ. 18-15. Valve assembly unit is mounted on the suction and discharge of a rotary-piston hydraulic elevator pump. Oil tank is on top of the steel enclosure.

3 to 7.5 hp, mounted on a steel frame floating on a vibration-absorbing mounting.

Valve Operation. The valve assembly (Fig. 18-14) performs eight separate functions. Automatic smooth-start valve *A* permits the oil to bypass when the pump starts. This valve gradually closes, diverting oil to the jack to start the elevator smoothly. High-pressure relief valve *H* guards against overloading the elevator mechanism from excessive pressures. The valve operates on a close pressure margin between the fully closed and opened positions. If the elevator is overloaded to the point at which the car cannot move, this valve opens, bypassing the full discharge into the tank.

On units with heavy sheaves and large motors, an automatic stopping valve, also at H, unloads the system after the car reaches the landing on the up-cycle. It compensates for flywheel action of sheaves and prevents excessive slide or up-drift of the car.

A tank shutoff valve T isolates the oil tank from the remainder of the equipment for quick repairs and servicing of the power unit. A check or oil-lock valve O closes when the car comes to rest and locks it in that position on the oil column. This valve closes quickly without any perceptible reverse flow of oil and without shock.

Landing Stops. An automatic leveling valve L ensures accurate landing stops within 0.25 in. under a wide load range. It gradually closes when the car is slightly above the landing, permitting the car to level back automatically at slow speed.

Main lowering valve M, hydraulically controlled through an electric pilot valve, opens gradually, giving a smooth down-start. The valve also closes gradually, bringing the car to a smooth stop. Manual-lowering handle F permits lowering the elevator if power fails or adjustments have to be made.

The equipments described give a general idea of the valve functions as applied to electrohydraulic control. An exception is the power unit made by the Shepard-Warner Elevator Company that uses diaphragm-operated accelerating and lowering valves. These are pilot-operated from jack pressure; a static pressure always exists because of weight of car and load.

The General Elevator Company uses up to three separate motor-driven pumps for smooth control of its electrohydraulic elevators. These pumps are controlled in sequence, starting and stopping the elevator smoothly in addition to making accurate landings.

Pumps. Many types of pumps have been used on electrohydraulic elevators, mostly of the rotary type. These include herringbone and spiral gear, multiple vane, rotating piston, screw, centrifugal, and other types. For elevator service the pump should operate quietly, have a discharge free of pulsations, and be efficient and reliable. Any mechanical noise developed by the pump and motor can be largely prevented from being transmitted to the elevator car by mounting the power unit on sound insulation and connecting the pump by a rubber hose to the valve assembly (Fig. 18-13).

Pulsations in the pump's discharge are transmitted through the fluid stream into the jack and from there to the car. These cannot be isolated, so the pumps used on elevators must be reasonably free of pulsations or the car will have an objectionable throb when in motion on passenger service. Several pumps have been developed that are satisfactory for electrohydraulic elevator service, all of them of the rotary type.

As might be expected, pumps that have satisfactory noise and pulsation levels for freight-elevator service may not be suitable for passenger application. These problems are generally something for the elevator and pump designers to worry about to produce a satisfactorily operating elevator.

Gear Pumps. These types have been used with spur, helical or spiral, and herringbone designs, but for elevator service the two latter are the most commonly used.

Figure 18-16 is a cross section through a spur-gear pump and shows how it operates. The gears rotate in the direction of the arrows and have very close clearance with the casing. Where the gear teeth contact they also should form a practically fluid-tight joint. These close clearances are necessary to keep leakage small from discharge to suction.

When the pump is primed, liquid flows in between the gear teeth as they separate on the suction side. This operation is similar to the way that liquid flows into the

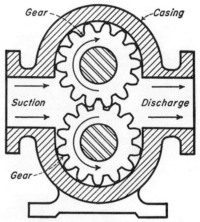

Fig. 18-16. Cross section through a spur-gear rotary pump.

cylinder of a reciprocating pump when the piston makes a suction stroke. As the gears rotate, fluid trapped between their teeth and the casing is carried around to the discharge. Since the fluid cannot escape back to the suction, sufficient pressure builds up to force it out the discharge pipe. Each gear tooth acts like the piston or plunger in a reciprocating pump to force liquid out the discharge.

One gear is fixed to the drive shaft and rotates with it while driving the idler gear, which may rotate on a fixed shaft or be fixed to a shaft carried in two bearings. Care has to be taken in designing the gear teeth so they will form a tight seal where they mesh and will not trap fluid in the root of the teeth to build up high pressure that will cause noisy operation and may overload the bearings.

Contrary to what is frequently accepted the flow from most designs of rotary pumps is not constant but slightly pulsating. This is be-

cause the fluid is delivered to the discharge in quantities equal to
the volume between the teeth. Because of this, spur-gear pumps
are not generally satisfactory for passenger electrohydraulic elevators. Objectionable features of spur-gear pumps have been eliminated to a high degree in herringbone- and helical-gear and other
designs of rotary pumps.

Herringbone-gear Pumps. Figure 18-17 shows a cross section
through a Worthington herringbone-gear pump that has been used
successfully in electrohydraulic elevator service. Its construction is
simple, with the gears and their shafts accurately located to rotate

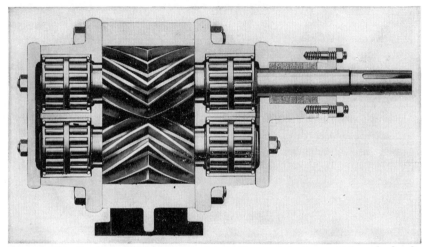

Fɪɢ. 18-17. Cross section through a herringbone-gear pump used on hydraulic
elevators.

within a fixed casing. Because of the herringbone shape of the gear
teeth axial forces produced between them as the gears mesh are
equal in both directions to neutralize the hydraulic thrust; therefore, a thrust bearing is not needed. Fluid cannot be trapped in
the gear teeth as they mesh to cause excessive loading of the bearings,
noise, and wear. The gear teeth gradually mesh and disengage to
provide for smooth filling and emptying of the spaces between them
so that hydraulic noises from this source are kept to a low level
when the pump is operated at medium speeds.

Because the two halves of herringbone-gear teeth form a wide vee
there may be a question as to the direction in which the pump should
rotate. They may rotate in either direction, but it is common practice to have the apexes of the teeth rotate toward each other. In

Fig. 18-17 this can be done by the teeth of the bottom gear moving up while those of the top gear move down, making this side of the pump the discharge.

In Fig. 18-17 the gear shafts run in cylindrical roller bearings that allow free end float of the gears and permit locating them near the load to keep shaft deflections to a minimum. A packing box is used

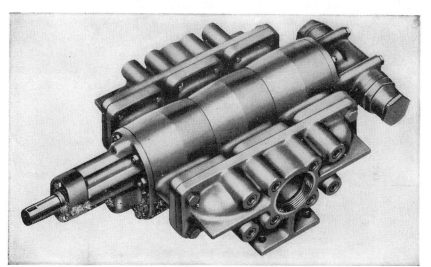

Fig. 18-18. Three-section helical-gear elevator pump with a relief valve on its right-hand end.

Fig. 18-19. Helical-gear rotors for the pump shown in Fig. 18-18 has each pair of gears separated by a bronze bearing.

on the pump where the drive shaft comes out of its bearing. However, where a small amount of leakage cannot be permitted the pump may be equipped with a mechanical seal.

Helical-gear Pumps. Figure 18-18 shows a pump of this type built by George D. Roper Corporation for elevator service, with a relief valve on its right-hand end. It is built in sections with pairs of equal-size gears that are assembled in units of one, two, or three sets of gears to obtain different capacities. In Fig. 18-18 the unit is a three-section assembly with three sets of gears, as in Fig. 18-19. All gears are free on their shafts. Special keys prevent drive shaft gears from turning on their shaft where the idler gears rotate on a fixed bronze shaft.

On the two- and three-section units the pairs of gears are separated by bronze bearing spacers which align the sections in the pump casing and provide additional bearings for the shafts. Ball bearings further support the drive shaft at each end and axial hydraulic thrust is also taken on a ball bearing. Pumps are rated at three different speeds, depending on required capacity and degree of quietness required. If noise is not a serious factor, pump speed can be 1,800 rpm; if a moderate amount of noise can be tolerated operation can be at 1,200 rpm. Where maximum quietness is needed

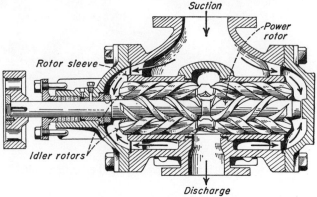

Fig. 18-20.   Screw pump with one power and two idle rotors that act as seals for the power element.

pumps are operated at 900 rpm and are double the size for a given capacity at 1,800 rpm. Quiet operation and efficiency is further improved by use of a venturi suction and discharge to reduce turbulence in the flow into and away from the pump. Action of the venturi causes the fluid to spread out across the full width of the gears and fill their tooth spaces. The shape of the venturi opening coordinates oil velocity to gear speed.

Screw Pumps. Figure 18-20 shows a DeLaval triple-screw pump used in electrohydraulic elevator service. It has a power rotor to which the power source connects directly. Two idler rotors act as seals for the power rotor and are driven by fluid pressure, not by metallic contact with the power rotor. Fluid coming in from the suction to the outer ends of the rotors is trapped in pockets formed by the threads and carried forward toward the discharge very much like a nut on the power screw. Handling oil, the pumps will operate at pressures up to 1,000 psi and at speeds up to 7,000 rpm. However, elevator service is usually at pressures of not over 400 psi and speeds of 1,800 rpm or less. Operation of this pump is such that it gives a discharge practically free of pulsations.

Rotating-piston Pumps. The Rotary Lift rotating-piston pump (Figs. 18-21 and 18-22) has three pistons of a cross section shown. Top and bottom pistons *A* and *B* do the pumping where *C* forms a seal between suction and discharge. All three rotating members are

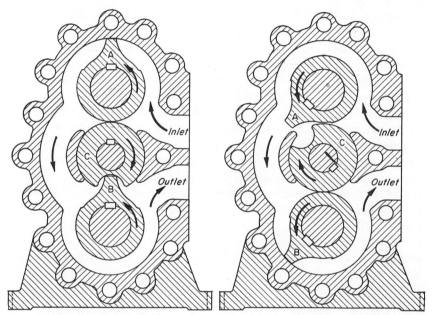

Fig. 18-21. Rotating-piston pump has three pistons, with piston *A* on discharge and suction.

Fig. 18-22. Same as Fig. 18-21, but with pistons rotated to where *B* is on discharge and suction.

geared together in a 1 to 1 ratio, so they rotate in step. There is no metallic contact of the rotating pistons with the housing. An oil film seals the unit. Because of the design of the rotors and their operation the pump's discharge is virtually free of pulsations.

In Fig. 18-21 rotor *A* is discharging and is also on suction. Oil flows freely in behind it through the suction line from an overhead reservoir. As piston *A* rotates, oil on its left is forced down and around piston *B* and out of the pump against discharge pressure. During this operation, even though *B* rotates it is idling and does not have to act as a seal, since pressure is equal around it.

In Fig. 18-22 the rotors have turned to the position where piston *B* is pumping against discharge pressure while piston A passes through its idling stage in oil at suction pressure. During the pumping stage, the back of rotating piston B rolls against sealing rotor C to form a seal between discharge and suction as between A and C in Fig. 18-21. During idling the rotating piston passes through the clearance notch in sealing rotor C, as shown in Fig. 18-21. There is no change in fluid velocity through the pump, as the oil passages are circular without abrupt turns, which contributes to smooth oil flow to the elevator jack.

Control Valves. Control of an electrohydraulic elevator involves the use of several valves, such as check, relief, bypass, lowering, and leveling. Check and relief valves are of conventional designs and serve their usual purposes of preventing reverse flow and overpressure in the hydraulic system. Bypass lowering and leveling valves are of special designs and will be considered here.

Bypass Valves. To start the elevator smoothly without shock and to reduce starting torque required by the pump motor a bypass starter valve is used. One design of these valves opens when the pump stops and closes when the pump starts. Others close when the pump stops, as in Fig. 18-23. When the pump starts, pressure below piston B immediately opens the valve to a position where pilot valve P closes the port through the center of the piston and holds it closed.

Pressure now increases above piston B through adjustable port A and the valve starts to close to increase fluid flow to the jack and accelerate the car. The rate of the valve closing, and consequently the rate at which car speed increases, is set by needle valve A, which is set to give a smooth rapid start. Pressure above pilot valve P holds it closed as the main valve moves down on its seat. Full operating pressure builds up above piston B to hold the valve on its seat. When the pump stops, pressure drops to where the spring under pilot valve P opens it to the position shown in Fig. 18-23 and the valve is again ready for the next start. The main valve can be operated manually by needle valve C.

Lowering and Leveling Valves. The car is raised by the pump discharging oil into the jack, but is lowered by a valve that opens and lets oil flow from the jack back to the oil-storage reservoir. One design of an electrically operated lowering valve by the J. D. Gould Company is shown closed in Fig. 18-24. When the main valve is closed, pilot valve P is also closed on its seat, so that full line pressure builds up above piston B. This pressure combined with compression of spring S holds the valve tightly closed. The rate at which pressure increases above piston *B* and the rate at which the valve closes are adjusted by needle valve *A*.

To open the valve, solenoid coil *D* is energized, which slams plunger *E* up the pilot valve stem against its top. The impact of the plunger on the pilot valve opens it and port *F* through the center of piston *B* to relieve pressure above it. Pilot orifice *F* being larger than port *A*, pressure above the piston drops to the point at which pressure below it can slowly open the valve.

As indicated by this valve's operation it is designed so that its rate

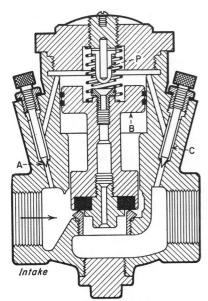

Fig. 18-23. Cross section through bypass valve used in hydraulic elevator control.

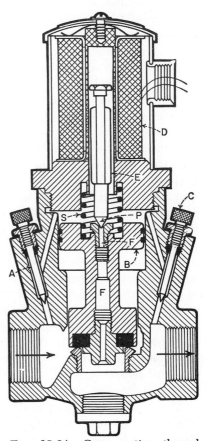

Fig. 18-24. Cross section through solenoid-operated lowering or leveling valve.

of opening or closing can be controlled. This is necessary for smooth down starting and stopping of the elevator. If the valve were opened quickly the car would accelerate downward too fast, and if it closed quickly the car would stop with a jolt. These conditions are unpleasant for the passengers and subject the equipment to unnecessary stresses and strains.

The valve shown in Fig. 18-24 in smaller sizes can be used also as a down leveling valve. Operation is the same as previously explained, but must be adjusted for quicker opening and closing.

In case of power failure, when the elevator car is at an upper floor it can be lowered by using the valve shown in Fig. 18-24 as a bypass, using needle valve C as a control. Opening this valve relieves pressure oil above piston B and the full system pressure opens the valve slowly to bypass oil from the jack back to the storage reservoir. When the car approaches the floor, closing needle valve C stops it. Car-lowering speed can be readily controlled by the position of the needle valve.

Elevator-valve Assembly Operation. Figure 18-25 is a simplified drawing of an electrohydraulic elevator valve assembly by Otis that combines all the valves for accurate smooth control of the elevator's operations. The valves in this assembly, even though they operate somewhat differently from those shown in Figs. 18-23 and 18-24, perform the same functions. The valves are shown in the position that they would be in before the elevator was started. Above the assembly is part of the oil-storage tank, with a connection A to the pump suction and B the return line from the elevator cylinder or jack.

The connection on the left goes to and from the jack, and below the assembly it connects to the pump's discharge. Bypass valve C is open and has a pressure closing port with a closing speed adjustment at D. Check valve E is closed. Lowering valve F and down leveling valve G are controlled by oil pressure through solenoid-actuated pilot valves H and J. The rate at which lowering valve F can open is controlled by the setting of needle valve K; its closing rate, by needle valve L. On the down leveling valve an opening speed adjustment only is provided by an adjustment at M.

The pump is a positive-displacement screw type especially designed for elevator operation. Assume that the elevator car is at the bottom landing and the pump is started by pressing a landing or car button. Part of its discharge goes through bypass valve C back to the storage reservoir and the rest through check valve E to the elevator jack starting the car in the up direction.

Oil pressure through orifice O is applied to the bottom of bypass valve C and it closes gradually, turning the full discharge into the jack and accelerating the car to full speed. When the car approaches the floor for which the button was pressed the pump is stopped as the car comes level with the landing. The car is oil-locked in position by the closing of check valve E. As the pressure drops around bypass valve C it opens, ready for the next start.

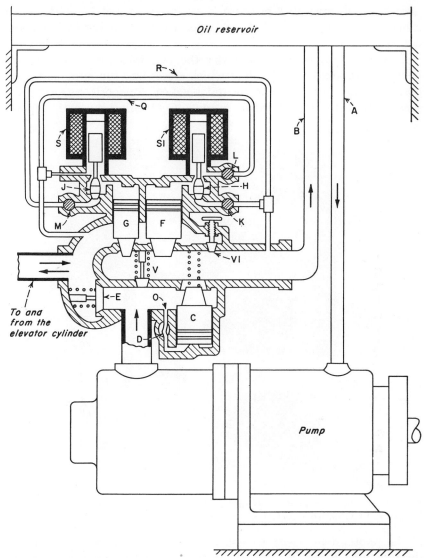

Fig. 18-25. Simplified drawing of a hydraulic valve assembly that combines all the valves for accurate and smooth control of an electrohydraulic elevator.

When the pump starts and builds up pressure in the line to the elevator jack, pilot valves *H* and *J* are in their down position with their top ports open and the bottom closed, as shown. Pressure oil is admitted by pipe *Q* through pilot valve *J* to the top of leveling valve *G* and through pilot valve *H* to the top of lowering valve *F* to hold these valves closed. Since pipe *Q* connects directly into the jack line, full operating pressure is on top of valves F and G at all times to hold them closed except when they are in use to lower the car.

Lowering Operations. To lower the car, solenoids S and S_1 are energized to lift the pilot valves which close their top ports and open their bottom ones. Closing the pilot valves' top ports cuts off pressure oil from the top of valves F and G. Opening the bottom ports

Fig. 18-26. Power unit includes motor M, pump P, and valve assembly D, with the oil reservoir above and electrical control on the left.

of the pilot valves opens a passage from the tops of valves F and G through pipe R into discharge line B to the oil reservoir. Valves G and F now have a low pressure above them and jack pressure below, and they slowly open. Their rate of opening is controlled by the setting of needle valves M and K, as previously explained. As valves F and G open, oil flows from the jack through them into the oil reservoir as the car starts downward. As the car approaches the desired floor it opens a switch in the down-valve solenoid circuit and its pilot valve closes. This in turn causes down valve F to start closing and slows the car down, and when it is near the floor the car opens a switch in the down leveling pilot-valve solenoid circuit and this valve closes. Leveling valve G closes quickly to bring the car to a soft nonjar stop at the floor.

These explanations cover valve operations for up and down travel of the car. If the car is overloaded to where the pump develops excessive oil pressure, relief valve V opens and bypasses oil to the storage reservoir. If power fails when the car is at an upper floor, it can be lowered with valve V_1. When this valve is opened it bypasses oil from the jack into the line leading into the storage reservoir. In this hydraulic control system the relief valve is part of the valve assembly and discharges into the oil reservoir. Here the oil is cooled in the large volume in the reservoir. On some valve arrangements the pressure-relief valve is on the pump and discharges into the suction of the pump. This has the disadvantage that only a small amount of oil is recirculated, which may cause it to overheat seriously in a short time, unless the high pressure is relieved or the pump stopped quickly.

The valves diagrammed in Fig. 18-25 are shown at D in Fig. 18-26 assembled in a compact unit, easily accessible for adjustment and maintenance. The assembly is factory-calibrated to give the desired characteristics for the elevator installation. Flexible hoses A and B connect the oil reservoir to pump P and the valve assembly D to the oil reservoir. Connections C from the valve assembly to the electrical control panel on the left are also flexible to reduce the transmission of noise and vibration to the hoistway and car. Connection between the valve assembly and elevator plunger cylinder is made at E and pump P is driven by motor M.

CHAPTER 19

Electrohydraulic Elevator Controls

Types. Controls for electrohydraulic elevators have many of the features found in those for electric types in comparable service. Electric controls are available to meet every need of low-rise passenger and freight electrohydraulic applications. For small light-duty freight elevators serving low-rises at low speed, a simple constant-pressure push-button control without automatic leveling might serve the purpose. Where the needs are more exacting or local codes must be met, controls that will hold the car level with the floors within $\frac{1}{4}$ in. are available. For passenger service up to four- or five-floors' rise and speeds of 100 to 125 fpm, momentary-pressure push-button control is generally used.

Electrohydraulic elevator controls may be divided into four general classes: 1, constant-pressure push-button operation; 2, momentary-pressure operation, commonly called single automatic operation; 3, nonselective-collective automatic or accumulative operation; and 4, selective-collective automatic operation.

Constant-pressure Push Button. With this system up and down buttons are provided in the car and at the landings, any one of which will control the movement of the car so long as it is held in the operating position. The elevator can be controlled from any floor as well as from within the car. This type of control is used chiefly for freight elevators where a simple type is adequate.

Momentary-pressure Operation. Commonly called single automatic control, it has a button in the car corresponding to each landing served and a single button at each floor. When any car or landing button is pressed, pressure on any other button has no effect on the car's operation until operation by the first button has been completed. This type of control is widely used for passenger or freight elevators handling medium traffic, such as in small hospitals and other buildings where elevator service is light.

338

Nonselective-collective Operation. This type of control functions automatically from one button in the car for each floor served and a single-call button at each landing. All stops registered by momentarily pressing a landing or car button are made independently of the number of buttons pressed or the sequence in which they are pressed. Car stops at all landings registered are made in the order in which they are reached, irrespective of direction of car travel.

Selective-collective Operation. This is one of the most advanced developments in hydraulic-elevator control. A button is provided in the car for each landing served and an up and a down button are placed at each intermediate landing with a call button at each terminal landing. All stops registered by the momentary pressing of buttons are answered automatically in the order in which the landings are reached in each direction of travel. Up-landing calls are answered when the car is traveling in the up direction and down-landing calls are answered when the car is in down direction. This control is suited for up to six-story office and apartment buildings where traffic flow is comparatively heavy in both directions.

Other Features. To these controls may be added many other features, such as automatic leveling, which functions to bring the car into a landing at reduced speed to ensure accurate and smooth stops. Automatic releveling can be provided to meet safety-code requirements. It holds the car level with the landing as long as the car is within the leveling zone. A noninterference feature prevents stealing the car by pressing another button until the car has completed its designated operation.

With or without an attendant, dual operation is desirable on elevators that carry peak traffic during the day and light traffic at night or on holidays. With this system the elevator controller is arranged for either automatic operation by the passengers or by an attendant in the car. When operated by an attendant, switches are thrown that prevent the car from being started from the landing buttons, which then are used to signal the car to the landings.

Control Operation. Figure 19-1 is a control wiring diagram for a Rotary Lift electrohydraulic elevator used in a one-story rise and is one of the simplest types. It is for a 5-hp 208-volt three-phase squirrel-cage motor driving the pump that supplies oil at 200 psi to the elevator jack. The motor is protected by thermal overload relays. Car and hoistway-gate contacts prevent unsafe operation of the elevator. These contacts are in series in the up- and the down-direction control circuits so that the car cannot start in either direction unless the gates are closed and locked.

Operation is by constant-pressure push buttons, one in the car for each direction of travel and one at each landing. As long as one of these buttons is held closed the car will continue in motion, until the button is released, or the car opens a floor limit switch. These switches, located in the hoistway near each floor, are in series with the up- and the down-direction contactor coils. As the car approaches a floor a cam on the former opens the limit switch, stopping the elevator as when the push button is released.

Assume that the car is at the bottom landing, the car and hoistway-gate contacts closed, and the up button in the car held closed. When the bottom power line is +, a circuit is completed from this

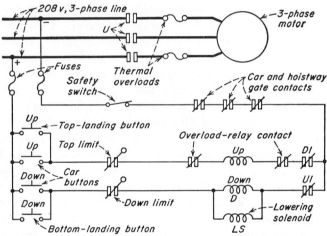

Fig. 19-1. Wiring diagram for a constant-pressure push-button control used on an electrohydraulic elevator for one-story rise.

line through the car up button, top limit switch, up contactor coil, overload-relay contacts, down solenoid contact D_1, gate contacts, the safety switch in the car, and to the − line.

Energizing the up-contactor coil closes contacts U in the motor circuit to drive the pump and lift the car with its load. As the car approaches the landing a cam on the former opens the top limit switch to stop the car at the floor, where it is held by the check valve in the oil line between the jack and pump. A smooth stop in the up direction is obtained by the energy stored in the rotating elements of the pump and motor and the time it takes the check valve to close. Up contactor U also opens contact U_1 in the down-coil circuits so that they cannot be energized from the down-landing button.

Down Operation. Once the car stops at the landing and the car and hoistway gates are opened it cannot be started from a landing or car button until the gates are closed again. When this is done, the car can be lowered by holding closed either the car or the landing down button. Assume that the bottom-landing button is closed. It completes a circuit through the down-limit switch, the down relay coil D, and lowering pilot-valve solenoid LS, contact U_1, and the gate contacts, to the — line.

Down relay D opens contact D_1 in the up-contactor coil circuit to isolate this circuit and prevent interference with down operation from the up buttons. Lowering solenoid LS opens the pilot valve that actuates the hydraulically controlled lowering valve. This valve is opened gradually, allowing the car to start down smoothly. It also closes gradually, bringing the car to a smooth stop when the push button is released or the down-limit switch opens.

If the car does not stop level with the landing it may be jogged level by the push buttons, as on an electric elevator. But cars on which this type of control is used operate at slow speed, 25 to 50 fpm, and are easily stopped level with the landings. They are also open so that anyone using the elevator can readily see when it approaches a landing to make a level stop.

Automatic Leveling. The control shown in Fig. 19-1 is about the simplest type, whereas that shown in Fig. 19-2 is a more complicated one. This control is for car speeds up to 100 fpm, and automatically levels at the landings. It also has a bell that sounds if a car or landing button is pressed when the car and hoistway gates are not closed and locked. Whereas the control shown in Fig. 19-1, has limit switches only to stop the car level at the top and bottom landings, that in Fig. 19-2 has anticreep and leveling switches to stop the car level with the floor and keep it there.

The top-stop and bottom-stop switches are in the hoistway near the floors and are opened by cams on the car to stop it level at the floors. The anticreep and leveling switches are on the car and are closed by cams in the hoistway to hold the car level at the floor. Because of leaking valves or packing, if the car creeps down from the floor it closes the anticreep switch, which returns it to a level position. On the other hand, when a stop is made above the floor a cam in the hoistway closes the leveling switch to energize the leveling valve solenoid L to open this valve. This is a small valve that bypasses the main down valve to bleed oil slowly from the jack and level the car with the floor. When the car is in this position the cam in the hoistway clears the leveling switch and it opens to let the leveling valve close.

Gate Circuits. A transformer supplies 115-volt power for the control circuits. Assume control-transformer polarity as shown, the car at the bottom landing, and the gate contacts open when the top-landing button is pressed. This forms a circuit from + on the transformer, through overload contacts *OL*, contact *SR*, top-landing up button, contact SR_2, and the bell transformer, to — on the con-

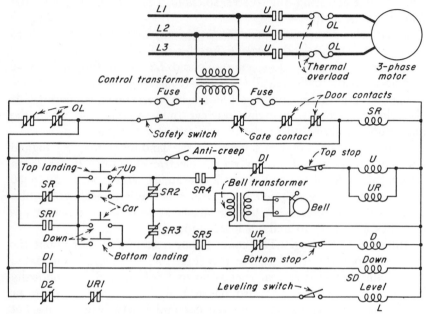

Fig. 19-2. Control wiring diagram for an electrohydraulic elevator that automatically levels the car with the landings.

trol transformer. This will sound the bell to show that the car gate or landing doors are not properly closed.

When the gate contacts are closed they form a circuit directly through signal-relay coil *SR*. This relay opens contact *SR* and closes SR_1 to connect the push buttons in series with the gate contacts. Relay *SR* also opens contacts SR_2 and SR_3 to isolate the signal bell from the push buttons and closes SR_4 and SR_5 in the up- and the down-coil circuits. If the up button in the car is held closed it completes a circuit from + on the transformer through the overload and gate contacts, contact SR_1, the car up push button, contacts SR_4 and D_1 top stop limit, up-contactor coil *U* and relay coil *UR* in parallel, to — on the transformer. Relay *UR* opens contact *UR* in the down-valve relay-coil circuit and contact UR_1 in the down-level switch circuit to prevent interference with the up-direction circuits.

Up-contactor coil *U* closes contacts *U* in the motor circuit and this motor drives the pump to supply oil to the jack to lift the car. As the car approaches the floor a cam on it opens the top stop switch to break the circuit through the up coil U and stop the car level with the floor. If the car creeps down from the floor a cam in the hoistway closes the anticreep switch to close the up-contactor-coil circuit and return the car level with the floor. If the car stops high at the floor a cam in the hoistway closes the leveling switch to lower the car to the floor.

Let us assume that we wish to lower the car; the gate contacts are closed and the down button is pressed at the bottom landing. This completes a circuit through down-relay coil D, which closes contact D_1 in the down solenoid-valve coil SD circuit. The valve opens to bleed oil from the jack back into the reservoir to lower the car to the bottom landing. This valve is of a design that gives the car a smooth down start. As the car nears the floor, cams open the bottom stop switch and closes the leveling switch so that as the down valve closes the leveling valve opens to bring the car to a smooth stop level with the floor. At this time the leveling switch opens and its valve closes. If the car creeps below the floor a cam closes the anticreep switch to start the pump and return the car to the floor.

Noninterference Relay. The control shown in Fig. 19-3 is similar to that shown in Fig. 19-2, except that its parts are not arranged in the same way, and there are a few more of them. Where the control in Fig. 19-2 has floor-stop and leveling switches only, that in Fig. 19-3 also has final-limit switches FL, that are opened by the car if it goes by a terminal floor. If the car opens one of these switches the elevator cannot be started again until the car has been moved off the switch by manual operation of the control. In the car the push buttons have three contacts, the middle one of which energizes the noninterference relay N.

When the final-limit switches and gate contacts are closed they complete a circuit through gate-relay coil DR that closes contacts DR and DR_1 and opens contacts DR_2 and DR_3. The latter two isolate the signal bell. When the car up button is closed its center contact completes a circuit through relay coil N. This relay opens contact N to isolate the landing buttons and closes contact N_1 to form a holding circuit for its coil. This relay cannot open until the car and hoistway gates open, so that even if a landing button were held closed it could not interfere with operation of the car even when it stops at the top floor with the gates closed. The gates must be opened and closed before the car can be started from a landing button.

Up-direction Operation. The right-hand contact of the car up button energizes up-relay coil *UA*, which closes contact *UA* in up-contactor coil *U* circuit. This coil closes contacts *U* in the pump-motor
circuit, and the car starts up and is stopped at the floor by the top
limit switch. If the car goes above the floor, a cam in the hoistway
closes leveling switch *LS* to energize leveling solenoid *LO*. This
solenoid opens a small leveling valve that lets oil flow slowly from

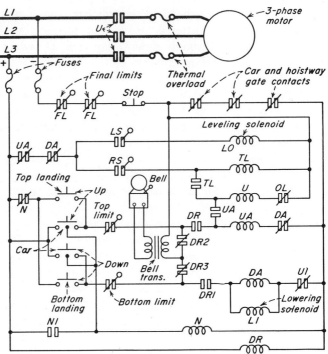

Fig. 19-3. Whereas the control in Fig. 19-2 has stop and leveling switches only,
this one has final-limit switches that are opened by the car if it goes by the
terminal landings.

the jack to bring the car to the floor, at which time leveling switch
LS opens. If the car goes below the floor it closes leveling switch
RS to energize relay *TL*, closing contact *TL* to energize up-contactor
coil *U*, and bringing the car up to the floor. Relay *TL* is of a timing design and will not close unless the car stops on switch *RS*.

When the car is not in use, the gate and final-limit contacts are
closed so that if the down-landing button is pressed a circuit is completed through contact *N*, the button, bottom limit switch, down-
relay coil *DA*, and lowering solenoid L_1, in parallel. Relay *DA*

opens contact *DA* in the up-relay coil *UA* circuit to prevent interference with the elevator's operation from the up button. The lowering solenoid opens the lowering valve to let oil flow from the jack to the reservoir to lower the car. Operation of this valve is the same as explained in Fig. 19-2 to bring the car level with the floor, where it is held level within ¼ in. by leveling switches *LS* and *RS*.

Momentary Push-button Control. The three controls described give a general picture of constant-pressure push-button types, without and with automatic leveling. Many of the controls used on electrohydraulic elevators, particularly for passenger service, are momentary push-button types. After a car or hall button is pressed and the initial circuit is completed the button can be released and the elevator continues to operate until stopped by a limit switch.

Figure 19-4 diagrams a Shepard-Warner control of this type with automatic power-operated car and hoistway doors, for a three-floor rise. It has many features of the constant-pressure push-button controls that we have studied and also others to adapt it to the company's elevator equipment and to keep the circuits closed after they have been completed by the pushing of a hall or car button.

Power lines, the three-phase 440-volt pump motor, and the 220-volt door-operator motor are at the top of the diagram. Also at the top of the diagram are the open-delta-connected transformers that supply 220-volt three-phase power to the door-operator motor and single-phase to the control circuits.

Start Valve. A solenoid pilot-actuated smooth-start valve connects across the motor terminals. This valve is normally open when the elevator stops but gradually closes when its solenoid coil is energized. Therefore, as soon as the pump starts the valve begins to close and increases oil flow to the jack to lift the elevator car. When it is completely closed full pump discharge goes to the jack and the car moves up at full speed. When a stop is initiated and the pump motor is cut off the line, the valve solenoid is deenergized and the valve opens. Consequently, if the pump continues to rotate, due to inertia of the pump, its motor, and drive, it does not cause the elevator to drift upward, because pump discharge is bypassed to the oil-storage reservoir.

Leveling Magnets. These are switching units that contain a powerful permanent horseshoe magnet, a permanent bar magnet, and an electrical contact. The magnets are mounted so that their like poles oppose, as shown in Fig. 19-5, to hold the contact open. These car-leveling units are mounted on the car and are actuated by steel vanes fastened to a guide rail at each floor.

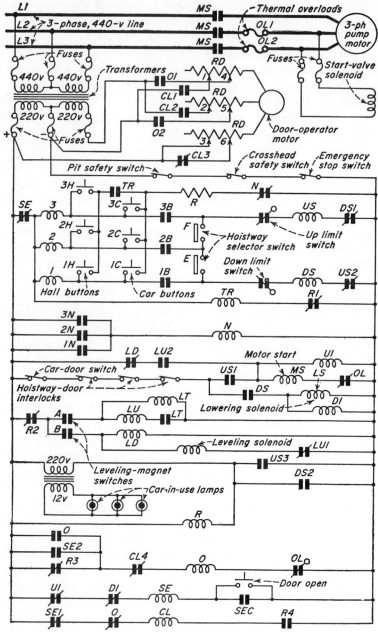

FIG. 19-4. Momentary-push-button control with automatic power-operated car and hoistway doors for a three-floor-rise electrohydraulic elevator.

When a vane comes between the magnets, as in Fig. 19-5, it provides a path for their flux. Therefore, the magnets no longer oppose but are attracted to the vane. The pull of the horseshoe magnet closes its contact to energize the leveling circuit. Consistent accurate floor leveling is obtained even though space between the vane and magnets may vary because of guide-shoe wear, or the use of roller guides.

In the control circuits are three safety switches, an emergency stop switch in the car, a car-crosshead safety switch, and a pit safety switch. If any one of these is open the car cannot start.

Control Operation Explained. Assume that the car is at the bottom landing and that the third-floor button $3C$ in the car is closed. This completes a circuit from $+$ on the transformers, through safety-edge relay contact SE, car-call relay coil 3, button $3C$, resistor R, control-relay contact N, and the safety switches, to $-$ on the transformers. Relay 3 closes contact $3B$, which completes a circuit for the up-direction switch, through the up-limit switch, direction-switch coil US, contact DS_1, and the safety switches, to $-$ on the transformers.

Closing contact $3B$ bypasses push button $3C$ so it can be released and direction-switch coil US remain energized. This switch opens contact US_2 in the down-direction-switch coil DS circuit to isolate this circuit while the car is in up service. Relay 3 closes contact $3N$ in the control-relay coil N circuit, which opens contact N in the push-button control-relay circuit to isolate these buttons from the start circuits.

Direction switch US closed contact US_3 in the run-indicator relay coil R circuit, which opens contact R_1 in the noninterference timer relay circuit (previously closed through contact SE), so that contact TR opens to prevent interference with car operation from the hall buttons. Contact R_2 also opened in the leveling circuits, so they cannot come into operation until the car is stopping at a floor. Contact R_3 opened in the door-opening circuit, so the doors cannot open when the elevator is in operation.

Car-in-use Lamps. Closing contact US_3 also energized the car-in-use lamps at the hall-call buttons. These lamps light if the car is in use from a car or hall button. They show anyone wishing to call the car from a hall button whether or not the car is in use. If the lamp at a hall button is dark it shows that the elevator is not in service. When a hall button is pressed and its lamp lights it shows that the elevator is responding to the call. If the lamp does not light, the waiting passenger knows that the elevator is not responding to the signal, such as might be the case if the hall and car doors are not properly closed.

Door-closing Operations. Closing contact R_4 in the door-closing-contactor coil CL circuit closes contacts CL_1 and CL_2 in the door-operator motor circuit and opens contact CL_3 in the door-opening circuit, and the car and hoistway door close. With contacts CL_1 and CL_2 closed and CL_3 open the transformers are connected to points 1, 2, and 3 in resistors RD. This puts a maximum resistance in series with the motor to close the doors at a safe speed. To open the doors contacts O_1, O_2, and CL_3 are closed to connect the transformer to points 4, 5, and 6 of the resistors. The motor now has reduced resistance in series with it and will open the doors faster than they are closed. This can be done safely because there is no danger of hitting a passenger when the doors are being opened, but there is when the doors are closing.

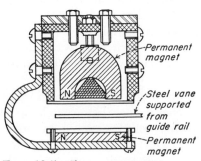

Permanent magnet

Steel vane supported from guide rail

Permanent magnet

Fig. 19-5. Permanent-magnet car-leveling unit.

The doors have a flexible rubber safety edge with contact SEC inside. When closing, if the doors press against a passenger the safety edge deflects and closes contact SEC to energize coil SE, which opens contact SE_1 in the door-closing circuit and closes contact SE_2 in the door-opening-contactor coil O circuit, and the doors immediately open. When the doors open, they open limit switch OL to deenergize opening-contactor coil O and disconnect the operator motor. As soon as pressure is relieved from the safety edge, contact SEC opens and this circuit returns to normal.

A door-open button is shown connected in parallel with safety-edge contact SEC. This button is on the car-control panel and serves to open the door when it is necessary and it is safe to do so. When the doors are open, limit switch OL cuts the power out of the operator motor, but when the doors close the operator motor remains connected. This motor is of the torque type that can remain connected to the power source permanently. With this type of door operator it is necessary to keep power on the operator motor to ensure that the doors do not coast open, as the operator does not have a self-locking gear train. However, when the elevator is not answering calls, relay R opens contact R_4 in the closing circuit to disconnect the door-closing circuit. As a result, power is applied to the door-operator motor only while the car is in actual service.

When the car and hoistway doors close they close their contacts to complete a circuit through direction pilot-relay coil U_1. Energizing this coil opens contact U_1 in the car safety-edge relay coil SE circuit so that the car and hoistway doors cannot be opened by pressing on the safety edge after the doors have closed.

Pump-motor Starting. Closing the car-door switch and hoistway-door interlock also completes a circuit for the pump-motor contactor coil MS through contact US_1, which closed when direction switch coil US was energized. Contacts MS close to energize the pump motor and the start-valve solenoid. The pump comes to speed quickly, but part of its initial discharge bypasses through the start valve, which closes automatically to accelerate the car smoothly, as previously explained. As the car reaches the top floor a cam on it opens the up limit switch to stop the pump motor and the start valve opens to stop the elevator.

When direction switch US opens it opens contact US_3 in run-indicator relay coil R circuit and contact R_3 closes in door-control relay coil O circuit. Contact R_4 also opens in the door-closing circuit and door-closing contactor CL opens to cut power out of the door-operator motor. Opening contactor CL closes contact CL_4 in the door-opening-contactor coil O circuit. This energizes coil O and close contacts O_1 and O_2 in the door-operator motor circuit to open the car and hoistway doors. As the doors approach the open position they open limit switch OL to disconnect the operator motor and the doors remain open.

Leveling the Car with the Landings. If the car does not stop level with the floor when direction switch US opens, opening contact US_3 in the run-director coil R circuit closes contact R_2 in the leveling circuits. When the car stops above the landing a steel plate in the hoistway causes the contact of leveling magnet switch B to close and energize coil LD of the down leveler and the leveling solenoid coil. Coil LD opens LD contact in the up-direction circuits U_1 and MS coils. Energizing the leveling solenoid opens the down-leveling valve to lower the car slowly to landing level, when contact B opens to stop the car and hold it at the landing. If the car stops below floor level, or oil leaks from the jack and lets the car creep down, contact A closes to complete a circuit through leveling timer coil LT, which closes contact LT in up-leveler coil LU circuit. This coil opens LU_1 contact in the leveler solenoid circuit to isolate it and closes contact LU_2 in the up-direction pilot-relay coil U_1 and motor-starting-contactor coil *MS* circuits. Up-direction pilot relay U_1 opens contact U_1 in the door safety-edge relay circuit to prevent the door from being opened by being pulled on the safety edge. Closing contacts *MS* starts the motor and pump to lift the car level with the floor when contact *A* opens to stop the pump.

Down Travel. Assume that the car is at the top landing when the second-floor hall button *2H* is closed, which completes a circuit through car-call relay 2 coil, and contacts *TR* and *N*, to — on the transformers. Car-call relay 2 closes contact *2B* for a circuit through selector-switch contact *E* which was closed by a cam on the car when in up direction to complete a circuit through down-switch coil *DS*. This switch opens contact DS_1 in the up-direction-switch coil circuit to isolate up operation and relay 2 closes contact *2N* in the control-relay *N* circuit to open contact *N* in up push-button circuits. Coil *DS* also closes contact DS_2 in the run-indicator coil *R* circuit to open contact R_1 in the noninterference timer *TR* circuit, R_2 in the leveling circuits, and R_3 in the door-opening circuit, and closes R_4 in the door-closing circuit to close the doors. Closing the doors closes their contacts to complete a circuit for the lowering solenoid direction-pilot relay coil D_1. The latter opens contact D_1 in the car-door safety-edge relay circuit to prevent interference with the doors after they have closed. Lowering solenoid *LS* opens the lowering valve to release oil from the jack to the storage tank as the car goes down to the second floor. As the car approaches this floor a cam on it opens selector switch *E* to stop the elevator and closes *F* to permit up travel of the car from the first floor. Opening switch *E* breaks the circuit through direction switch coil *DS* that opens contact *DS* in the lowering-valve solenoid *LS* to allow this valve to close and stop the car at the floor. A series of operations are initiated similar to those explained for up motion stopping to open the car and hoistway doors and stop the car level with the floor.

When the car is at the second floor, pressing car button *1C* or hall button *1H* will complete a circuit through car-call relay coil 1 and contact *N* to close contact *1B* in the down-direction switch coil *DS* to initiate down operation of the car. Similarly, closing car button *3C* or hall button *3H* starts the car in the up direction. Other features of this control could be covered in more detail, but sufficient instruction has been given to lead to a complete understanding of its many functions.

CHAPTER 20

Escalators and Moving Sidewalks

People have become so convenience-conscious that they may object to walking up or down one flight of stairs. For this and other reasons a need has developed for more and better low-rise vertical transportation in office buildings, department stores, theatres, railway terminals, subway and elevated stations, factories, and other public places. For these services, what may be classed as moving stairways are used in large numbers. They are built under four different names: Otis, escalators; Westinghouse, electric stairways; Peelle, motorstairs; and Haughton, moving stairs. All operate on the same general principle, varying chiefly in design and arrangement of parts. In The American Standard Safety Code for Elevators they are all classed as escalators, as they will be in this chapter.

Supporting Truss. These machines are really a highly refined conveyor system. They have three main parts: 1, two traveling handrails; 2, moving stairs; and 3, a power unit, usually an electric motor and a speed-reducing gear. Figure 20-1 shows these parts in their correct relation to each other, supported within a steel truss, shown in Fig. 20-2.

Factory-built of standard steel shapes with track brackets welded in place, the truss eliminates difficult field erection operations and cuts installation costs. It is rigidly constructed and consists of a top-end section, a long straight center section, and a bottom-end section, welded together on the job to form one continuous structure.

Ample strength is ensured in all directions by horizontal and diagonal braces. An oil pan is welded inside the bottom of the truss along its full length to catch any oil drips from the moving stairway. The truss is supported from the building steel to form a rigid structure in which to assemble the stairway and hold it in alignment.

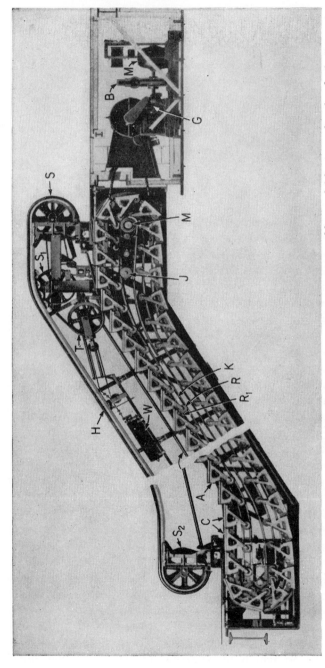

Fig. 20-1.   Cross section through an escalator, which has three main parts: two rubber and fabric traveling handrails H, moving stairs assembled in a steel truss, and a power unit, usually an electric motor M and a speed-reducing gear.

Traveling Handrails. In Fig. 20-1 the traveling handrails *H* are of rubber and fabric construction and travel on a guide on top of the balustrade. At their top end each passes down over a sheave *S* and back into the balustrade, where it passes around a tension sheave *T* and a fixed sheave S_1. From this sheave, travel is down through guides and around a fixed sheave S_2 at the lower end of the stairway.

Each tension sheave is mounted on a guide frame and connects by a wire rope that passes over two small sheaves to a weight *W*. This weight keeps correct tension in the handrail if its length increases from use. When this stretch approaches the limit of the tension device, a switch closes to light a lamp in the motor room. If the handrail is not shortened, a second limit switch opens and stops the stairway when the limit of stretch is reached or the handrail breaks.

To drive the handrails a roller chain connects a sprocket on the shafts of fixed sheaves *S* and S_1 to a sprocket on a jack shaft *J*. A sprocket on this shaft connects by a chain to a sprocket on the main drive shaft *M*.

Handrail Construction. In Fig. 20-1 the handrails are made of rubber and duck specially formed to operate on a track *G* (Fig. 20-3). Joints are vulcanized to provide a continuous smooth rubber surface. The metal track is shaped to the inside of the handrail to keep friction low and insure cool operation. The design of the handrails and their supporting guides is as important to safety as it is to maintenance. The handrails fit around their guide rails to form a safety grip without a finger-pinching hazard. There are no ledges or obstructions against which fingers might be pressed or pinched.

Openings into which the handrails enter the balustrade at the landings are placed near the floor under the casings around the sheaves. These openings are so out of the way as to make it practically impossible to get fingers pinched at these points. A stiff brush, rubber, or felt guard around the handrails at these openings gives added protection.

A handrail design similar to that shown in Fig. 20-3 has a thin steel tape embedded in its inner surface, which practically eliminates stretch and the use of automatic tightening devices. Tension is maintained in this design by a simple manually adjusted idler sheave behind the driven one in the upper newels. The handrails are vulcanized endless of the correct length in the shop, to eliminate a difficult job in the field.

Figure 20-4 shows part of a metal handrail that consists of sections of closely coiled nonstaining aluminum springs, shown black and white in the photograph. Clamps *C* hold the sections together and connect them to a roller chain R and to laminated linen and plastic shoes S that run between lubricated steel guides. At the top and bottom landings the chain runs on sprockets and is driven from the lower step-chain sprocket shaft.

Step Arrangements. Step design is one of the most important features in successful escalator operation. As might be expected, step design follows approximately the same pattern in all makes, the

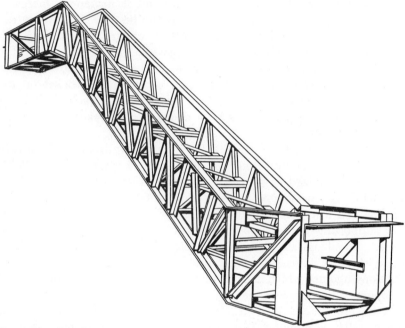

Fig. 20-2. Rigidly constructed structural-steel truss in which the moving stairway is assembled.

chief differences in them being in details of construction rather than in the general design. One step design is constructed as shown in Fig. 20-5, and is carried on two chain drives, shown in Fig. 20-6. Constructed with hardened and ground steel rollers the chains are quietly transferred from and to the track systems as the chains pass around their drive sprocket.

The inside edge of each step connects at A to the shaft running through the two chains, one on each side of the steps. On their riser side, the steps are supported on a shaft and two wheels. These wheels are shown at R in Fig. 20-1, and run on a track at K. Wheels R_1 are on the shaft passing through the chain (Fig. 20-6) and sup-

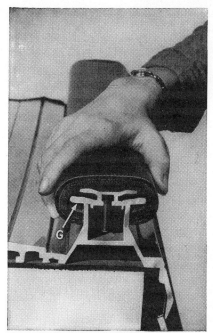

FIG. 20-3. Cross section through a rubber-and-fabric handrail specially formed to run on a track G.

FIG. 20-4. Metal handrail that consists of sections of closely coiled stainless-aluminum springs.

FIG. 20-5. Escalator step has a metal-cleat tread carried on four ball-bearing wheels, two on the step and one on each chain.

port the steps and chains on their tracks so that both are carried clear of all stationary parts to ensure quiet operation. A rail above the chain wheels prevents up-thrust of the chains and holds the steps in correct alignment.

In Fig. 20-1 the steel flanged rollers of the step-drive chains mesh with the drive sprockets. Here the step-drive chains transmit power only, the step load being taken by the step wheels. A later design of low-rise medium-capacity escalator has chain rollers and step wheels the same size and made of the same material. In this design every third drive-chain roller is a step wheel and all rollers run

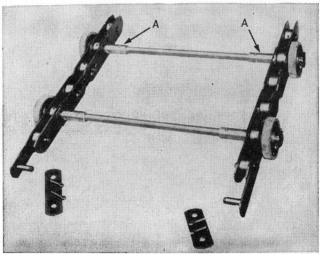

Fig. 20-6. A section of the two-step chains that drive the steps by connections from their inside edge to the wheel shaft at A.

on heat-treated aluminum-alloy tracks. As double-duty members the chain wheels support the load as well as transmit power, with the step axles also serving as chain pins. The load is distributed among many wheels to ensure long service from them.

Figure 20-7 shows one end of three steps and the drive chain with the step treads at S and the risers R between them. C and C_1 are chain rollers with C_1 serving as chain roller and step wheel with the step axle at A and the step trailer wheel at T. Step risers are of black enamel on sheet metal instead of the stainless steel used on many stairways. Black enamel is used to reduce glare and to outline the steps better.

Step and Chain Wheels. Large nonmetallic step and chain wheels running on antifriction bearings contribute much to quiet operation of modern escalators. One design of wheel has a tire made of canvas segments on edge bonded to a plastic center. Other wheels have silent duck treads bonded to micarta centers. Others have rubber tires, and still others have a tread made of compressed and impregnated canvas cut on the bias.

All wheels are constructed to operate with a minimum of noise and to act as sound insulation between the steps and tracks. They have a wide rolling surface that tends to prevent sliding and forming flat spots on them. All wheels run on double ball bearings, some

Fig. 20-7. One end of three steps with their trailer rollers and drive chain. Step risers are black enamel on sheet metal instead of the stainless steel used in many stairways.

of which have grease-gun lubrication and others sealed-in lubrication, assumed to be good for the life of the bearing.

How Driven. On all escalators the step chains run on sprockets at their top landings, from which they are driven (Fig. 20-1). Drive sprockets D are shown in Fig. 20-8 for one type of power end. In this unit the power-driving roller chain runs on sprockets P and the handrail-drive roller chains run on sprockets H, one for each rail. Sprocket G drives the speed governor. A magnetic brake between sprockets P and D holds the stairway from running free when power is off.

In Fig. 20-8 the sprockets are fixed on the axle and all revolve as a unit, whereas in Fig. 20-1 the axle is stationary and the step-chain sprockets revolve on it on roller bearings (Fig. 20-9). Step-chain sprockets D are rigidly connected together by a cast-steel spider, or barrel housing BH, to which they are bolted to prevent misalignment. In Fig. 20-8 the handrails are driven by sprockets H at each end of the main drive shaft, but in Fig. 20-9 these sprockets are on the barrel housing at H between the step-chain sprockets.

Again in Fig. 20-8 the speed governor is driven by a sprocket G on the main drive shaft, whereas in Fig. 20-1 it is at G on the power unit and chain driven from a sprocket on the worm-gear shaft. The opposite end of the worm-gear shaft connects by a double-strand

FIG. 20-8. Drive end of a moving stairway with the step-chain, power, and handrail- and governor-drive sprockets mounted directly on the drive shaft.

roller chain to a sprocket P on the stairway sprocket-drive assembly (Fig. 20-8). The governor can be driven with safety from the power unit since if the main drive chain breaks a broken-chain safety switch opens the power circuit and the stairway is stopped by brakes B and B_1 on the step-chain sprocket (Fig. 20-9).

Stairway Brakes. On the driving machine shown in Fig. 20-9, there are two pawl-applied separately operated brakes side by side on the right-hand step-chain sprocket. These have two pawl rings B and B_1 faced with friction materials F and held between plates O by tension springs S. As long as stairway operation is normal, two magnets hold the pawls released from rings B and B_1, so that the whole brake rotates with the drive-chain sprockets.

When power is cut off by the opening of a stop or safety switch,

one magnet releases the pawl into one of the pawl rings. This stops
the ring, which slows the stairway down because of the braking action; but the other ring continues to rotate with the drive shaft. A
few seconds later the second magnet releases its pawl, which applies
the second brake to bring the stairway to a smooth stop. In this
system both brakes operate each time the stairway stops, at least
once a day. When there is no service brake on the power unit the
brakes on the step-chain drive shaft make both service and emergency stops. When there is a service brake on the power unit, as at
B in Fig. 20-1, then the brakes on the step-chain drive shaft are used

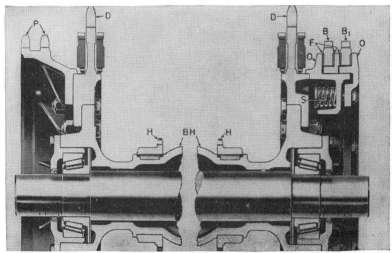

Fig. 20-9. On this escalator drive end the step-chain, power, and handrail-drive
sprockets are on a cast-steel barrel housing.

for emergency stops only and are applied by a broken drive-chain
switch.

Terminal Sprockets and Tracks. At the top end of the stairway
shown in Fig. 20-1, step chains are guided around the sprockets by
curved rails so that the step wheels follow quietly with the chains.
At the lower landing, the step and chain wheels are guided in curved
tracks without sprockets, with the steps following around as at the
top landing. As the steps come around the track carriage they run
level a short distance, as shown at C in Fig. 20-1, to permit easy
landing; then they gradually rise to their full height, as at A.

At the lower terminal the stairway carriage and tracks are mounted
on guide wheels, carried on a short horizontal track for the carriage
to move and keep proper tension in the step chains. In Fig. 20-1

this tension is maintained by springs. Any slight wear or stretch of the chain is taken up automatically by the springs moving the carriage to the left. Movement of the carriage is limited to about 2 in. when a safety switch opens to stop the stairs. This switch also acts as a broken-chain safety device to stop the stairway if either chain break or operates in an abnormal manner. Other escalator designs have lower-end step-chain sprockets to help guide the steps.

Fig. 20-10. Stairway power unit has a squirrel-cage motor *M* and a brake *B*, both supported from the lower half of the worm-and-gear housing, a rugged one-piece casting.

Power Drives. Escalators are driven at constant speed by a power unit that consists of a constant-speed motor and some form of speed reducer. In Fig. 20-1 the power unit has a single-speed squirrel-cage motor *M*, a service brake *B*, and a worm gear, and is similar to the power unit of a worm-gear elevator machine. This machine is also built without a service brake; the brake on the step-chain sprocket acts for both service and emergency stops.

Other escalators are driven by a closely assembled motor, worm gear, and brake of either horizontal or vertical construction (Figs. 20-10 and 20-11). The unit shown in Fig. 20-10 has a squirrel-cage motor *M*, and a brake *B* supported from the lower half of the worm-gear housing, a rugged one-piece casting to give rigid support to the whole machine. An overspread and underspread governor *G* is driven directly from the end of the motor shaft. The gear and sprocket shaft rotate on heavy-duty roller bearings, and the worm thrust is taken by a double-row ball bearing. A broken-chain switch is at *C*.

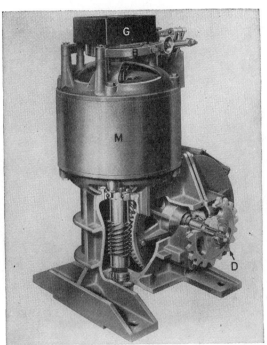

Fig. 20-11. Power unit similar to that shown in Fig. 20-10, but in a vertical assembly for small-size stairways.

The power unit shown in Fig. 20-11 is similar to that in Fig. 20-10, except that it is of a vertical assembly for driving smaller escalators than that in Fig. 20-10. The power unit in Fig. 20-11 is designed for installation in the top end of the truss, as at *M* in Fig. 20-12, to save space and reduce building framing. A heavy single-strand roller chain connects the drive sprocket *D* on the machine to that on the step-chain drive shaft. Since the service brake is on the power unit (Figs. 20-10 and 20-11) and a chain drive is used, an emergency brake is provided on the drive shaft.

In still another design, the power unit has an axial-air-gap squirrel-cage motor *M* mounted on a helical-gear speed reducer *S* con-

nected to the step-chain drive shaft by a triple-strand roller chain
C (Fig. 20-13). There is no brake on this unit; the safety brake *B*
on the step-chain drive shaft, applied by weight *W* on a lever,
serves as both service and emergency brake. The brake stops the
moving steps if power fails to the motor or if any of the safety
switches open. Braking action is adjusted so that the stairway is
stopped in about a 2-ft run. This brings the steps to rest without

Fig. 20-12. Power unit shown in Fig. 20-11, installed at *M* in the top end of the
escalator truss to reduce space.

throwing the passengers off balance. The brake is released only in
starting sequence initiated by a key-operated starting switch.

Safety Features. These stairways are the safest means of transportation in use even though they are subject to all the pranks that
human ingenuity, curiosity, and carelessness can devise. They are
made safe for even the most careless people to ride on. This is because safety has been designed into every part of the equipment, beginning with the steps, which have narrow treads of wear-resistant
aluminum alloy that form a stable platform, comfortable to stand on.

Space between the cleats on the steps is so narrow that soft soles or narrow high heels cannot sink between them. Close clearances prevent dragging of galoshes or boots between the treads and skirt risers. Solid step risers and close clearances prevent accidents as the step formation changes.

On the old type of flat-step moving stairways with smooth tread steps a shunt S (Fig. 20-14) was required at the landings to guide the passengers on and off the steps. At the loading end of modern

Fig. 20-13. This escalator power unit has an axial-air-gap squirrel-cage motor M mounted on a helical-gear speed reducer S connected to the step-chain drive shaft by a triple-strand roller chain.

stairways the steps come out from under a comb, as shown at C in Fig. 20-15. The comb teeth fit with close clearance between the step-cleat treads to form practically a smooth surface. Distinct safety features are the ample level spaces formed at the landings by the two step treads T nearest the comb plate. When in this level position the two adjacent treads fit so closely together that there is no space or groove in which to snag a heel or catch the point of an umbrella.

At the unloading end of the stairway the comb teeth remove small objects that may lodge between the step cleats and also slide passengers' feet from the moving steps onto the comb plate if they forget to step off. On most modern escalators combs are made of a special semibrittle aluminum alloy with the teeth designed to snap off before bending. Experience shows that it is safer to have the teeth break than to have them bend upward and form a tripping hazard. If a tooth breaks, as it infrequently does, a repair can be quickly made by replacing the damaged comb section.

On some stairways, to distinguish more clearly between the step-tread cleats and comb teeth, the comb plate and teeth are made of

Fig. 20-14. Old type of smooth-tread-step escalator required a shunt S at the landings to guide passengers on and off.

black molded thermosetting plastic. Except for the difference in color these combs have about the same characteristics as those made of aluminum and operate in a similar manner.

Extended Newels. To get the passengers on and off an escalator is one of the most important problems in its safe operation. Semicircular newels (Fig. 20-15) at both landings contribute much to solving this problem. In Fig. 20-15 the newel and balustrade extend about the width of two steps before the steps start to rise. For part of this distance the steps are under the floor and comb plates; then they come out and run flat for about two tread widths before they start to rise. In this distance, passengers can grasp the moving handrails before reaching the steps and retain their grip until after they leave them. This assists them to enter and leave the stairs smoothly and without hesitating, and practically eliminates the stumbling hazard on the steps.

Extending the newels also allows placing the openings *H* into which the handrails disappear into the balustrades, where there is little chance of finger-pinching accidents. Finger guards are placed

Fɪɢ. 20-15. On modern escalators the comb teeth fit closely between the step cleats to form a practically smooth surface at the landings.

around the moving handrails where they enter the balustrades. One design of guard closes the clearances around the rails with a brush. Even if someone were to get down near the floor and stick his finger into the brush it could be easily removed without pinching. On other moving stairways the openings around the handrails are protected by rubber and felt guards.

Just above the comb plate at each terminal a light *L* in the side of the balustrade (Fig. 20-15) illuminates the landing plate and comb for added safety for those getting on and off the stairway. Escalators are well illuminated for their full length.

Safety Stops. In addition to the features that make it safe for passengers to ride on escalators other safety devices stop the stairway if it is not in good operating condition. These include: 1, an emergency stop button at each landing in a readily accessible position; 2, a broken or slack step-chain device in the lower carriage; 3, slack or broken handrail switches; 4, overspeed and underspeed governor contacts; 5, on chain-driven installations a broken or slack drive-chain switch; and 6, on some jobs, a comb-plate and floor-plate switch; these are arranged so that if one of them opens it cuts power out of the driving motor and applies a brake to stop the stairway.

Besides these safeties the motor control has an overload relay to protect the motor, and on some a reverse-phase relay to prevent the motor from being reversed by the crossing of the power leads outside the motor room.

Fire Hazards. Elevator hoistways and escalator runs form open wells through which smoke, gas, fumes, and flames can escape from one floor to another in case of fire. Complete enclosure of elevator hoistways with fire-resistant material is generally considered adequate protection against fire, if the top of the hoistway is properly vented and the hoistway doors are solid.

Escalators offer a different problem because they normally run open. In stores there may be serious objections to enclosing escalator runs. It is while customers are riding on the escalators that merchandise can best be displayed to them.

When open runs may be a fire hazard there are several ways to provide protection. One of these totally encloses the runs with fire-resisting material and requires doors at each landing. These doors are generally open but close automatically by fusible links or other means in case of fire. Even though this method gives a high degree of protection and is one of the simplest to install, it has serious disadvantages. It is more difficult for people to find the escalators than it is when they are open and in stores customers cannot see merchandise while riding. Also, it is pleasanter to ride on an open stairway than on one which is enclosed.

A second method uses a fire-resistant smoke guard F around the wellway opening at each floor (Fig. 20-16). This guard extends about 18 in. down from the ceiling. Sprinkler heads H are located about 4 ft apart around the guards.

A method similar to the smoke guard uses nozzles to produce a high-velocity water spray around the wellway opening instead of the usual sprinkler heads. A deep fire-spray water curtain or barrier is produced which gives an effective counteraction to smoke.

Roller-type shutters S, made of fire-resisting material, operating on a railing R about 3 ft high at the upper landing have been used for fire protection (Fig. 20-16). The shutters may be closed by hand or by power. When power-operated, closing may be initiated by heat, smoke, or other conditions. An 18-in. smoke guard is usually used

Fig. 20-16. A fire-resistant smoke guard with sprinkler heads around the well-way opening at each floor and roller-type fire-resistant shutters operating on a railing at the upper landings have been used for fire protection.

with this method. In Fig. 20-16 a smoke guard F, sprinkler heads H, and roller shutters S are used for fire protection.

One of the most effective methods is the sprinkler-vent system, even though it is not generally used. It consists of an exhaust system separate from the normal building ventilation, with suitable ducts, flues, and dampers to draw out air, smoke, and gases in case of fire, and an intake through which fresh air enters the building. When it is in operation a down-draft is created through the wellway and gas and smoke are exhausted from each floor affected.

Operation of the system is initiated by a thermostat on the floor in trouble or by devices as required for automatic fire protection. Also, there are smoke guard and a water-spray system around the stairway openings, controlled by thermostatic devices spaced to respond to advancing heat. Excessive heat will automatically create an effective water curtain from the ceiling to the floor.

These systems will provide adequate fire protection for practically every moving-stairway installation. Which one will serve a given set of conditions best has to be decided by the architect, the consulting engineer, or escalator manufacturer to meet the code requirements of a given city.

Capacity. For handling large crowds, escalators have many advantages, the first being their large capacity. A single escalator 4-ft wide and operating at 90 fpm will handle up to 8,000 passengers per hr, whereas a large low-rise elevator can serve only about 500. A moving stairway handles as many passengers on a 70 ft rise as on 12 ft. Height of rise is a limiting factor in the number of passengers that an elevator can handle.

An escalator loses no time in loading, starting, stopping, and unloading, and there is no crowding in and out as on elevator cars. You can step on an escalator at any time and be carried up or down at a moderate rate, comfortably, quietly, safely, and without physical effort other than walking on and off. No operator is required and power consumption is low.

Escalators are not limited to handling large crowds. Frequently they are installed for interoffice service, where one company occupies more than one floor. When they are used to connect the ground and second floors of an office building, the latter is almost as desirable for renting as the former. In department stores they increase sales volume on all off-street floors by making them easily accessible and reducing customer fatigue by eliminating waiting and crowding. These and many other factors have made escalators as popular as elevators for low- and medium-rise service in many types of buildings.

Rise Limits. Escalators have the disadvantage of being limited in the rise for which they can be built, but gradually this limit is being increased. Two units in the Lexington Avenue and 53rd Street station of the Independent Subway, New York, have a vertical lift of 56 ft. On the subway system in London, England, there are 18 escalators serving vertical rises of 58 to 81 ft. Some of these have capacities of up to 16,000 passengers per hr and operate at speeds of up to 180 fpm.

Escalator speeds in the United States have, in general, been limited to 90 fpm, run at a 30° angle to the horizontal. However, many have been installed to operate at speeds of 120 fpm.

Higher Operating Speeds. Escalator capacity does not increase in proportion to speed. Increasing the speed 33 per cent from 90 to 120 fpm increases capacity only about 20 per cent. This is because of the difficulty some people have in getting on an escalator. They hesitate and this leaves open spaces in loading, which reduces the hourly capacity. In a few cases speed has been reduced from 90 to 75 fpm to improve loading. This was done with little or no loss in capacity. On these escalators loading conditions apparently were not originally as good as they might have been.

However, there is a tendency toward increasing speed from 90 to 120 fpm, permitted by most city codes. There seems to be no good reason for not doing this when the approach to the escalator is properly designed.

Two-speed Operation. Escalators are being installed with two-speed motors to operate them at 90 or 120 fpm. On some of these installations the lower speed was provided for use if 120 fpm proved to be too fast for the conditions. However, it was found that passengers loaded as readily at the higher speed as at the lower. As a result the escalators are operated at the higher speed all the time.

Two-speed escalators have been installed to operate at high speed during peak-load periods and at low speed during off peaks. Other systems of two-speed operation have been proposed but two-speed operation has not settled into standard patterns.

Escalator Widths. Standard stairways are now 32 and 48 in. wide between the moving handrails, but have been built 24, 32, 36, and 48 in. wide. Step lengths are about 8 in. less than stair widths. The 32-in. width was introduced to meet the demand for a low-rise unit in small stores and other places. This width will carry a parent and child comfortably on the same step and will accommodate a shopper with her bundles. Passenger capacities at 90 fpm are: 48 in., 8,000; 36 in., 6,000; 32 in., 5,000; and 24 in., 4,000.

Escalators in Series. Even though they are limited in the height of a single rise for which they can be built, escalators are operated in series to serve several floors as is often done in large department stores. Operation in this way requires a transfer at each floor, which is not a serious objection in stores, as a large part of the traffic is from floor to floor. Escalators can be operated ascending or descending or can be installed to operate in either direction. To accommodate traffic peaks they may ascend for part of the day and descend at other times.

Higher Rises. To adapt escalators to higher buildings a system combining express and local stairways was proposed by G. B. Gusrae in the April, 1952, *Architectural Record.* In a ten-story building the express units would serve the first, fourth, seventh, and tenth floors (Fig. 20-17). Thus, a passenger could go from the first to the tenth floor and transfer only twice, compared to eight times on the local. Travel time on the express is 2 min, 31 sec; and on the local, 3 min, 19 sec. To go from the first to the sixth floor a passenger would express to the fourth and take the local to the sixth, requiring two transfers and 2 min, 47 sec to make the trip.

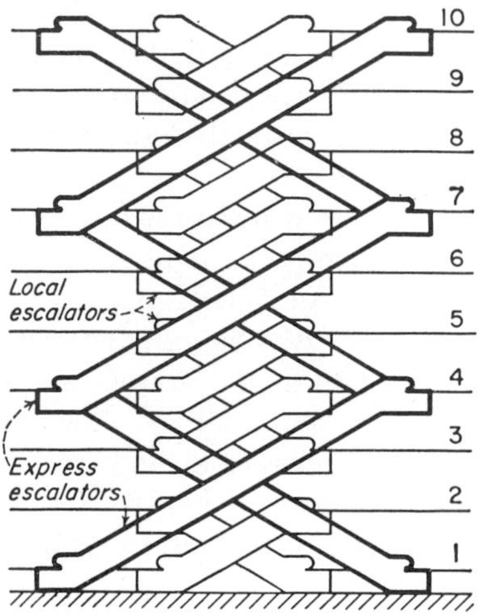

Mr. Gusrae arrived at the conclusion that besides lower daily operating cost, escalators do not require penthouses, pits, or complicated controls, and they need considerably less floor space than elevators of comparable capacity and cost less. Breakdowns affect only one unit at a time, with the parallel unit available as a spare for upward travel. The equipment is simpler and easier to maintain. Its life is equal to or longer than that of elevators. The system described using express and local moving escalators may easily prove acceptable for buildings up to twenty stories high.

FIG. 20-17. A proposed system of express and local moving stairways in a 10-story building.

Drive-motor Controls. An escalator is a slow-moving medium-friction medium-inertia constant-speed load; therefore it can be driven by a normal starting-torque motor. When alternating current is available, and it generally is, the motor may be a single-speed across-the-line start squirrel-cage type. Three-phase squirrel-cage motors with all six winding leads brought for star-delta starting are used extensively, as is primary-resistance starting on the larger escalators.

Compared to elevator-motor controls, those for escalators are simple. They are usually of a push-button-start type, arranged to reverse the power unit readily when necessary and actuate an electric brake. A feature of this control is the large number of safety devices that protect passengers and equipment against almost every operating hazard.

For the larger-size escalator, Westinghouse may use a primary-resistance start control (Fig. 20-18) with the power unit shown in Fig. 20-10. This machine is similar to that used on a horizontal-shaft worm-and-gear driver elevator except that it is designed for escalator service. At the top of the diagram (Fig. 20-18) are the up- and down-direction switch contacts U_1, U_2 and D_1, D_2. Below these are the thermal overload relays OL_1 and OL_2 and the potential switch contacts P_1 and P_2. Then come the three-phase motor windings M_1, M_2, and M_3 that connect to starting resistors R_1, R_2, and R_3, cut out of circuit by closing contacts SP_1 and SP_2.

Connected to the three-phase line is a reverse-phase relay RP and two control-circuit transformers grouped open-delta. These transformers have six sets of terminals brought out on them (not shown) that permit them to be operated on 199–208 to 550–575 volt circuits and supply the correct voltage for the three-phase copper oxide rectifiers, E_1 to E_6, that supply direct current to the brake-coil and control circuits. Assume transformer polarity as shown; then current flows down from $+$ on the two transformers, through rectifiers E_2 and E_6, and through the right-hand fuse. Current flow continues through the control circuits back to and up through rectifier E_3, and to $-$ between the two transformers.

Assuming that the center tap is $+$ and the two outside transformer terminals $-$, then current flow is from the center tap through rectifier E_4, the right-hand fuse, and the control circuits as before, through rectifiers E_1 and E_5 to the transformers. Even though current flows reversed in the transformers the rectifiers kept it in the same direction in the control and brake circuits.

Below the rectifier is the timing relay with coils T and T_1 that controls the time when resistors R_1 to R_3 are cut out of the motor circuits. Coils T and T_1 of this relay are connected directly across the rectifiers. As long as both coils are energized contact T_2 in speed-relay coil SP circuit remains open.

Brake Coil. Next comes the brake-coil circuit. Resistor R_6 shunts this coil so that when its circuit is opened to apply the brake the coil discharges through this resistor. This action causes the brake to be applied easily to bring the stairway to rest smoothly and without shock. Resistor R_7 is shunted by closed contact B_1, which connects the brake coil directly across the line for a quick positive release of the brake. When the brake releases, contact B_1 opens to put resistor R_7 in series with the brake coil to reduce the current in it to a holding value.

Parallel with the brake-coil circuit is full-speed contactor coil SP that closes contacts SP_1 and SP_2 to cut starting resistors R_1 to R_3 out

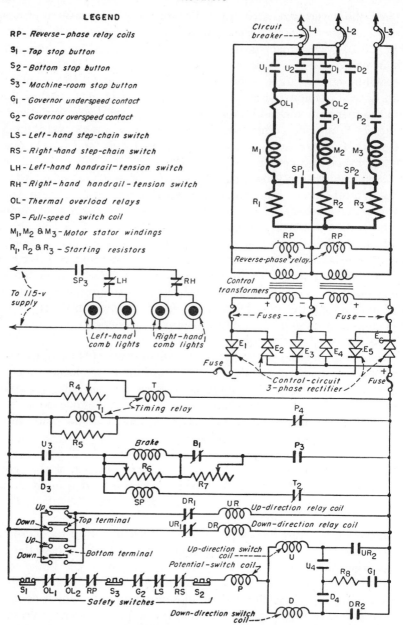

LEGEND

RP– *Reverse–phase relay coils*

S_1 – *Top stop button*

S_2 – *Bottom stop button*

S_3 – *Machine-room stop button*

G_1 – *Governor underspeed contact*

G_2 – *Governor overspeed contact*

LS – *Left–hand step-chain switch*

RS – *Right-hand step-chain switch*

LH – *Left-hand handrail–tension switch*

RH – *Right–hand handrail–tension switch*

OL – *Thermal overload relays*

SP – *Full-speed switch coil*

$M_1, M_2 \& M_3$ – *Motor stator windings*

$R_1, R_2 \& R_3$ – *Starting resistors*

FIG. 20-18. Control wiring diagram for a three-phase squirrel-cage motor with primary-resistance starting for driving a large escalator that can be reversed.

of the motor circuits for full-speed operation. Closing of these contacts is controlled by timing-relay contact T_2, as previously mentioned. At the bottom of Fig. 20-18 are the stairway terminal starting buttons, the direction-switch coils, and the safety switches.

Four step-comb lights, two for each terminal, are shown at the center of Fig. 20-18, separate from the control circuits. These lamps light the approach to the steps at each terminal, operate from a 115-volt power supply, and are controlled by contact SP_3 on the full-speed switch SP.

Control Operation. Assume that the stairway is to be started in the up direction from the top landing. At this station the starting key is put in the switch and turned to the up-start position, which closes the top up button. A circuit is completed from $+$ on the rectifier, through coil UR, contact DR_1, and the top up button, to $-$ on the rectifier. Contact UR_1 opens in the down-direction relay coil DR circuit to lock this relay open. Relay UR also closes contact UR_2 in the up-direction switch coil U and potential-switch coil P circuit. This circuit is from $+$ on the rectifier, through contact UR_2, coils U and P in series, and the safety switches, to $-$ on the rectifier.

Several contacts close and others open at the same time. Contacts P_3 and U_3 close in the brake-coil circuit to release the brake and let the motor start the stairway. Contacts U_1, U_2, P_1, and P_2 close in the motor circuits. When lines L_1 and L_3 are $+$ and L_2 is $-$, motor circuits are from L_1 through contacts U_1, winding M_1, resistors R_1 and R_2, winding M_2, and contacts P_1 and U_2, to L_2. The other motor circuit is from L_3, through P_2, winding M_3, and resistors R_3 and R_2, to L_2 as before. The motor starts with resistors R_1, R_2, and R_3 in series with its stator windings.

Coil P opens contact P_4 in the timing-relay coil T_1 circuit and this relay starts timing out. After a given time the relay closes contact T_2 in the accelerating coil SP circuit. This coil closes contacts SP_1 and SP_2 between resistors R_1, R_2, and R_3, to cut these resistors out by accelerating-switch coil SP. In series with each pair of lights is a of the motor circuit and bring it and the stairway to full speed. When the stairway reaches about 80 per cent of its speed the governor's underspeed contact G_1 closes. This completes a circuit for the direction- and potential-switch coils U and P through contact U_4 that closed when the U contactor coil was energized. The circuit is now directly from $+$ on the rectifier, through contact G_1, resistor R_8, contact U_4, coils U and P, and the safety switches, to $-$ on the rectifiers. The start button can now be released and the escalator remain in operation until one of the stop buttons or safety switches is opened.

On this control the comb lights are on a separate 115-volt alternating current. These lamps are lighted when contact SP_3 is closed by accelerating-switch coil SP. In series with each pair of lights is a slack-handrail contact LH and RH. If either of the handrails gets slack its switch opens to put out the comb lights and show that the handrail needs attention.

To start the escalator down from the top landing, the starting key is put in the control switch and turned to down-start position, which closes the top down button. A circuit is now completed from $+$ on the rectifier, through coil DR, contact UR_1, and the top down button, to $-$ on the rectifier. Contact DR_1 opens in the up-direction-relay coil UR circuit to isolate the relay. Relay coil DR also closes contact DR_2 in the down-direction switch coil D and potential-switch coil P circuit. From here control operation is practically the same as explained for up direction, except that contacts D_1 and D_2 close in the motor circuit to reverse the direction of motor rotation and travel of the escalator.

A description of this diagram gives a general idea of escalator-motor controls and the elements assembled in them. Even though these controls are quite simple they vary in circuit arrangements and safety features, but an understanding of Fig. 20-18 will greatly assist when studying the circuits of the others.

Moving Sidewalks and Ramps. For safe mass transportation of people over distances of up to 300 ft or more, horizontally or on inclines of up to 15°, what is generally known as moving sidewalks and ramps have been developed. They are used in rail, bus, and air terminals, shopping centers, parking lots, sports arenas, race tracks, under and over street crossings, through passageways, past exhibits in private and public buildings, and in factory traffic lanes.

These are available in two general designs. One, the Travolator by Otis, is a modified escalator with short metal platforms instead of steps. These platforms form a smooth flat surface for safe riding. They are cleated as are escalator steps with the cleat ends for one platform meshing with those of the two adjacent ones, as shown at *A* in Fig. 20-19. The cleated platforms are linked together by a chain drive to form a continuous surface, which travels on a wheel and track system laid out to suit the contour requirements of the installation.

Fixed landings at entrance and exit points are horizontal, making it easy for passengers to get on and off the travel strip. To reduce tripping hazards as riders step off the travel strip, each landing is edged with a comb plate, as on an escalator, shown at *A* in Fig. 20-20.

Extended newels with traveling handrails make it easy and safe for passengers to enter or leave the travel-strip.

Travolators are driven by a power unit similar to that of an escalator and have a single-action braking system for smooth stopping under light or heavy loads. Other safety devices include a broken-chain safety switch, a nonreversing mechanism, controller overload relay, and an overspeed governor. Much that has been said about

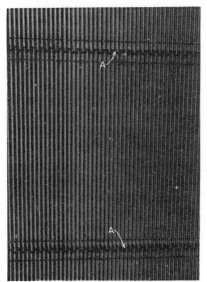

FIG. 20-19. Cleat ends on one platform mesh with those on the two adjacent ones, as at A.

FIG. 20-20. To reduce the tripping hazard as riders step off or on the travel strip each landing has a comb plate A.

escalators in this chapter applies to Travolators, including their reversibility.

These equipments are built in two widths. A 48-in. design on which two adults can ride side by side without touching elbows has a capacity of 12,000 passengers per hr. A 32-in. width, which can comfortably accommodate an adult and child standing side by side, can handle 7,200 passengers per hr. These capacities are based on a travel speed of 135 fpm either horizontally or on inclines of up to 15°.

Speedwalks and Speedramps. Another design of moving sidewalks and ramps is that by Stephens-Adamson, known as Speedwalks and Speedramps. These are a simple form of flat-endless-belt conveyor which can travel in either direction to suit the prevailing flow of pedestrian traffic, either horizontally or on inclines of up to 15° at a speed of 120 fpm.

They are built in roller and slider types. With roller supports, closely spaced ball-bearing rollers (Fig. 20-21) carry the belt. This type requires about 50 to 80 per cent less power than the slider-type installation and is used for heavy traffic loads and for belts that travel 300 ft or more. On the slider-type support (Fig. 20-22), the belt slides freely over a flat steel supporting deck on a special low-friction composition surface, practical for runs of up to 300 ft.

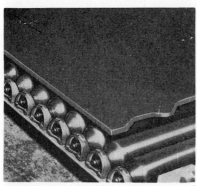

FIG. 20-21. On the roller type of belt supports, closely spaced ball-bearing rollers carry the belt to reduce required driving power.

FIG. 20-22. With a slider type of belt support, the belt glides freely over a steel deck on a special low-friction composition surface.

Belts are flanked on each side with a balustrade of either of three types. For heavy duty and on slopes an enclosed balustrade with a traveling handrail, like that on an escalator, is used. In normal service an enclosed balustrade with stationary handrail is adequate. An open balustrade with stationary handrail serves where cost is a factor.

Belt arrangements and their drives are quite similar to those for belt conveyors used in material-handling service. They are available in five belt widths, 24 to 108 in., with capacities ranging from 3,600 to 18,000 passengers per hr. On long installations they may be arranged for intermediate side exits and entrances as well as terminal loading and unloading.

Link-Belt also supplied information on flat-belt moving sidewalks and ramps.

Index